Springer Theses

Recognizing Outstanding Ph.D. Research

Aims and Scope

The series "Springer Theses" brings together a selection of the very best Ph.D. theses from around the world and across the physical sciences. Each volume has been selected for its scientific excellence by an internationally renowned institute or department. For greater accessibility to non-specialists, the published versions include an extended introduction, as well as a foreword by the student's supervisor explaining the special relevance of the work for the field. The series as a whole provides a valuable resource both for newcomers to the research fields, and for other scientists seeking detailed background information on special questions. Finally, it serves as an accredited documentation of the valuable contributions made by today's younger generation of scientists.

How to Apply

Theses prepared at internationally leading universities and institutes may be nominated for publication by supervisors with the endorsement of the head of department. Each such department is entitled to nominate one thesis per year. Enquiries and requests for a nomination form should be sent, together with the thesis, to Ms. Nathalie Berg (see link on right).

Eligibility Criteria

To qualify for consideration, theses should fulfil the following criteria:

- They must be written in good English.
- The topic should fall within the confines of Chemistry, Physics, Earth Sciences, Engineering and related interdisciplinary fields such as Materials, Nanoscience, Chemical Engineering, Complex Systems and Biophysics.
- The work reported in the thesis must represent a significant scientific advance.
- A maximum of 30% of the text may be a verbatim reproduction from the author's previous publications.
- They must have been examined and passed during the 12 months prior to nomination.
- The theses should have a clearly defined structure including an introduction accessible to new PhD students and scientists not expert in the relevant field.
- After acceptance, theses should be supplemented by a "Supervisor's Foreword" outlining the wider scientific relevance of the work.

Indexed by zbMATH.

Florian Stäbler

Dissipation, Interactions, and Heat Transport in Quantum Hall Devices

Non-equilibrium Bosonization of Quasi-One-Dimensional Systems

Doctoral Thesis accepted by
Université de Genève, Genève, Switzerland

Springer

Author
Dr. Florian Stäbler
Département de Physique Théorique
Université de Genève
Genève, Switzerland

Supervisor
Prof. Eugene V. Sukhorukov
Département de Physique Théorique
Université de Genève
Genève, Switzerland

ISSN 2190-5053 ISSN 2190-5061 (electronic)
Springer Theses
ISBN 978-3-031-90670-1 ISBN 978-3-031-90671-8 (eBook)
https://doi.org/10.1007/978-3-031-90671-8

This Springer imprint is published by the registered company Springer Nature Switzerland AG
The registered company address is: Gewerbestrasse 11, 6330 Cham, Switzerland

Supervisor's Foreword

The study of quantum transport in low-dimensional systems has long been a cornerstone of condensed matter physics, revealing fundamental aspects of electron interactions, dissipation, and quantum coherence. The thesis of Dr. Florian Stäbler represents a significant contribution to this field, addressing a central unresolved issue: the theory of dissipation in chiral and non-chiral quasi-one-dimensional systems.

One of the primary motivations for this research was the long-standing "missing heat" paradox in quantum Hall systems, an experimental puzzle that has challenged conventional theoretical frameworks. To address this problem, Florian focused on the role of mesoscopic Ohmic contacts, microscopic metallic islands coupled to quantum Hall edge states, which serve as key elements in quantum transport experiments. Through this exploration, the thesis not only sheds light on the missing heat paradox but also uncovers a wealth of new physical phenomena across four major research projects.

The development of a transmission line model to address energy dissipation in quantum Hall systems provides insight into the missing heat paradox. The study of nonlocal interactions in drift-diffusion systems, revealing how they can bypass equilibrium constraints and enhance heat transfer through interaction-induced correlations. The introduction of self-looping edge states on Ohmic contacts leads to novel transport properties, including anomalous correlation functions and modified response coefficients. The application of a Langevin-like approach to analyze the effects of the heat Coulomb blockade on heat noise power and temperature fluctuations, demonstrating the emergence of non-Gaussian Full Counting Statistics in charge transport.

Beyond addressing fundamental theoretical questions, Florian's research makes concrete predictions that can be tested experimentally, offering new directions for the fields of electronics miniaturization, precision metrology, and quantum information science.

Among the most remarkable achievements of this work is the prediction of the heat-multiplication effect, where strong electron correlations in self-looping channels lead to heat flux values exceeding the expected quantum limit. Another significant contribution is the theoretical framework for studying heat transport in non-chiral transmission lines, culminating in the discovery of a giant heat flux carried by chiral edge currents—an effect that remains to be fully explored both theoretically and experimentally. Additionally, the thesis provides a refined understanding of energy dissipation in quantum Hall systems, clarifying the role of disorder, interactions, and neutral excitations in restoring energy flux quantization even under strong dissipation.

The methodological rigor of this work is equally impressive. Using a diverse set of analytical tools, including bosonization techniques, Langevin approaches, and transmission line modeling, Florian presents a unified perspective on dissipation and energy transport in mesoscopic systems. This research exemplifies the power of theoretical physics to unravel complex experimental puzzles and lays a solid foundation for future advancements in mesoscopic and quantum transport physics.

I am pleased to see this outstanding Ph.D. thesis recognized in the Springer Thesis Series, a testament to its originality, depth, and impact. The work of Florian represents an important step forward in our understanding of quantum transport phenomena, and I am confident that it will serve as a valuable resource for researchers working in this fascinating field.

Genève, Switzerland
February 2025

Prof. Eugene V. Sukhorukov

Acknowledgements This thesis represents a personal milestone, not only in terms of my academic journey and the intellectual challenge to obtain a Ph.D., but also in terms of the people I encountered and the friends I made on this path. This part of the thesis is dedicated to these people.

First and foremost, I extend my deepest gratitude to my Prof. Eugene V. Sukhorukov. His passion for physics, enriching discussions, and friendly demeanor were not only a driving force behind my motivation but also led to the fantastic results summarized in this thesis. I am also grateful to Prof. Thierry Giamarchi (University of Geneva), Prof. Dr. Thomas Ihn (ETH Zürich), and Prof. Frédéric Pierre (C2N) for their roles in evaluating this thesis as members of my dissertation committee.

To Isi, my partner in this wild and sometimes downright crazy adventure of a multi-year long-distance love, I thank you for being my constant. Your love, patience and unwavering belief in me have been my anchor in the storm and my biggest motivation to continue to push forward. Every challenge has felt lighter with you by my side (even from miles away), and every triumph has been sweeter because I get to share it with you. I could not have done this without you—thank you for being my home, no matter the distance.

Moving to a new country can be tough in the beginning, which is why words fall short to express my immense appreciation for Michael Sonner—affectionately known as "Michi 1", a tremendous climber, cook, party host, Rhône descender, and our very own portable ChatGPT. His warm and welcoming nature was crucial in introducing me to the, at the time small, Ph.D. group—a lifeboat in complicated times. These individuals have become among my closest friends, him included.

Among them is Andreas, the boy with the golden palate, who turned everything Plainpalais offered on a Sunday into fermented delicacies, making him one of the best cook I've ever known. Our friendship extends beyond the kitchen to deep conversations, memorable hikes, countless laughs over my Instagram reel dumps, and competitive yet hilarious rounds of Cards Against Humanity. To Lucia, the Karaoke black belt and creative spirit of our group. An occasional climber, but a regular pizza enthusiast and founding member of the "Pizzaclimbers", she knows not just every episode of "Friends" by heart but also how to be one in the deepest sense. To João, whose creativity and passion for board games are matched only by his aversion for vegetables—a sentiment that, I'm pleased to note, has been improving. Our journeys together, particularly our travels along the US East Coast, are among my favourite memories. Don't miss out on checking out $|\mathrm{Hop}\rangle$. To Ralphy, gamer and YouTuber, known for his rapid-fire speech. It's with him that I share valuable memories, particularly from the early stages of my Ph.D. journey.

A special mention must go to my friend Julian, whom I've had the pleasure of knowing for almost ten years now. We found ourselves pursuing Ph.Ds. in the same field, in the same city and shared neighboring offices for nearly as long as we've known each other. Our time together has been filled with memorable evenings of wine, exquisite food, music, meaningful conversations, sometimes just silence, along with hikes, "pissuntersboot", vacations, almost relationship-ending "random" encounters at the Jet d'Eau and much more adventures. Here's to hoping we can continue this journey together.

I'm extremely grateful to the aforementioned "Pizzaclimbers" for countless indoor and outdoor adventures, climbs, hikes, ski trips, and the constant shop talk, especially during lunch. To Dario, Marko, Jerome, Elisabetta, Martino, Ivo, and also including Christophe Berthod, who introduced us to the stunning beauty of Valais in both summer and winter.

One of the most influential aspects of my time in Geneva was undoubtedly the sheer number of Italians I was fortunate enough to meet. They passionately upheld their Italian culture, particularly when it came to matters of food. Special thanks to Ludo, Lorenzo, and Giulia T., especially for the Italian support during ski trips; to Francesco, who is most memorable for his amazing Mascarpone chicken recipe—though he's now a vegetarian, leaving me to prepare it myself; to Giulia V. and Margherita; and a special mention to Giacomo, who not only fits into many of the categories previously mentioned but also assisted me in writing the French résumé of my thesis.

Speaking of Italian influence, this is the perfect moment to express my gratitude towards my colocs at "Casa della Gazzosa," which was my home for nearly the entirety of my Ph.D. I hope I managed to dissipate any initial skepticism and prove that Germans can indeed make for pleasant company. I hope I succeeded in upholding the house's spirit while also imprinting my own mark, particularly through the modernization of the meme wall. Special thanks to Diego, who went to great lengths to procure Guanciale for me from Italy; to Simone, who taught me the art of preparing it; to Davide, whose intellectual, philosophical, and moral discussions were not only inspirational but also a great source of enjoyment. Davide also played a pivotal role in guiding me towards the path I aim to follow post-Ph.D. Lastly, I extend my thanks to Michael Straub—affectionately, and somewhat involuntarily, dubbed "Michi 2".

I'd like to highlight the warm and friendly environment I found at the University of Geneva, particularly within the lunch group, composed of Dima's (old) group and Thierry's group. Special thanks to the crew of postdocs, some lucky enough to hit the academic jackpot with permanent positions. They were amazing companions through and through. I am grateful to Johannes, Alessio, Louk, Tony, Alexios, Michele, Nirvana, Catalin, Ilya, Alex, and Willem for the unforgettable moments, road trips, parties, and valuable discussions we shared.

I'd also like to extend my gratitude to my friends from Munich, with whom I've managed to stay in touch, and some of whom have even visited me, Christiane, Vroni, Benni, Ralph, Rudi, Anna, Stefan, Manu, as well as to my family, Anja, Uwe and Sebi.

Lastly, on a scientific note, I wish to acknowledge the fruitful discussions and the hospitality of Christian Spånslätt and Janine Splettstoesser, with whom I collaborated on a project and who helped improving this thesis. I also extend my gratitude to Mikhail Kiselev and Anton Parafilo for the fruitful discussions regarding my work.

Competing Interests The author has no competing interests to declare that are relevant to the content of this manuscript.

Publications Related to This Thesis

I. **Transmission line approach to transport of heat in chiral systems with dissipation**
Florian Stäbler and Eugene V. Sukhorukov
Phys. Rev. B 105, 235417 (2022)

II. **Non-local transport of heat in equilibrium drift-diffusion systems**
Florian Stäbler and Eugene V. Sukhorukov
Phys. Rev. B 107, 045403 (2023)

III. **Mesoscopic heat multiplier and fractionalizer**
Florian Stäbler and Eugene V. Sukhorukov
Phys. Rev. B 108, 235405 (2023)

IV. **Impact of potential and temperature fluctuations on charge and heat transport in quantum Hall edges in the heat Coulomb blockade regime**
Christian Spånslätt, Florian Stäbler, Eugene V. Sukhorukov and Janine Splettstoesser
Phys. Rev. B 110, 075431 (2024)

Contents

Acronyms

1D	One-dimensional
2DEG	Two-dimensional electron gas
AG	Aleiner-Glazman (mode)
(F)QHE	(Fractional) Quantum Hall Effect
OC	Ohmic Contact
QPC	Quantum Point contact
TL	Transmission Line

Part I
Introduction and Theoretical Foundations

Chapter 1
Introduction

Abstract Mesoscopic physics and quantum Hall systems provide a compelling intersection of experimental accessibility and deep theoretical insight, making them essential for exploring low-dimensional quantum phenomena. Mesoscopic systems, which exist between the macroscopic and quantum scales, reveal striking quantum effects due to electron confinement in small geometries. This confinement gives rise to discrete energy levels, quantized conductance, and coherence phenomena, offering a crucial bridge between classical and quantum mechanics. A key example of these quantum effects is the Quantum Hall Effect, observed in two-dimensional electron gases subjected to strong magnetic fields. The Quantum Hall effect, characterized by precisely quantized Hall resistance, has not only validated fundamental quantum principles but also revolutionized our understanding of electronic systems through topological concepts. These findings highlight the fundamental role of mesoscopic physics in advancing quantum science and developing future quantum technologies.

1.1 Mesoscopic Systems and the Quantum Hall Effect

Mesoscopic physics and quantum Hall systems stand out as platforms that beautifully blend experimental accessibility with rich theoretical frameworks, providing insight into the captivating realm of low-dimensional physics.

The physics of mesoscopic systems, characterized by their scale lying between the macroscopic world and the quantum realm, offer a unique vantage point. It allows us to probe the transition from classical to quantum behavior, where quantum effects become pronounced because of the confinement of electronic systems in small geometries. This confinement leads to discrete energy levels, quantized conductance, and the emergence of coherence phenomena that are otherwise obscured in the macroscopic world. It is this scale that bridges our everyday experiences with the fundamentally discrete nature of quantum mechanics, making mesoscopic systems a cornerstone for advancing our understanding of quantum phenomena.

The Quantum Hall Effect (QHE), discovered in two-dimensional electron gas (2DEG) systems subjected to strong magnetic fields [1], is a prime example of the

F. Stäbler, *Dissipation, Interactions, and Heat Transport in Quantum Hall Devices*, Springer Theses, https://doi.org/10.1007/978-3-031-90671-8_1

exotic phenomena that arise in low-dimensional systems. The QHE not only confirmed the quantization of Hall resistance, which is precisely determined by fundamental constants, but also introduced a new paradigm for understanding electronic systems in terms of topology [2]. This discovery was seminal not only because it provided a deeper understanding of electron dynamics in two dimensions but also because of its implications for precision metrology and potential applications in quantum computing.

In this Introduction, we focus on two types of seminal experiments. Their results and the open questions they produced have served as the basis for the projects explored in this book. These experiments provide a suitable context, setting the stage for the ideas and methods that define our research, which makes them a good starting point for this thesis. The first set of experiments focuses on the Ohmic contact, while the second set delves into energy relaxation in quantum Hall edge states. Throughout this book, we adopt a unit system in which the fundamental constants e, $\hbar$, and k_B are set to 1.

1.2 The Ohmic Contact

The Ohmic contact (OC), a small piece of metal placed on a highly doped semiconductor region containing a two-dimensional electron gas, is coupled to one or several one-dimensional (1D) chiral and ballistic quantum Hall edge channels, see Fig. 1.1. The contact is considered floating in terms of charge and energy conservation. A key property of this system is the single channel contact "RC" time[1] $\tau_C \equiv R_q C$, where $R_q \equiv 2\pi/e^2$ is the quantum of resistance and C is the contact capacitance. The RC-time is the charge relaxation time and is directly related to the contact charging energy $E_C \equiv e^2/(2C) = \pi/\tau_C$. At low temperatures, the interplay between chiral charge transport and finite RC time enables the Ohmic contact to act as a highly sensitive bandpass filter for transport. This property has been used to explore a wide array of intriguing quantum phenomena, such as dynamical Coulomb blockade [3, 4] in and out of equilibrium, heat Coulomb blockade [5, 6], electron state teleportation [7, 8], Coulomb-mediated heat drag [9], Luttinger liquid physics [10–12], charge fractionalization [13, 14], tunable multichannel Kondo effects [15–21]. Moreover, the OC provides a low resistance contact between edge states and external circuits, allowed the control of dephasing in interference experiments [22–30] and offers tunable energy relaxation [31–33] through tunneling, establishing it as an invaluable component in contemporary electron optic experiments. Additionally, it serves as an incoherent source and a detector of quasi-particles. A "real" electron micrograph of a typical device can be found in [5].

[1] For N channels, it would be $\tau_{C,N} = R_q C/N$.

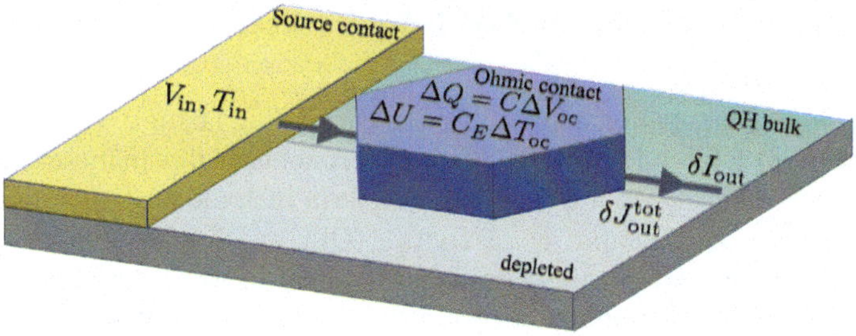

Fig. 1.1 Schematic representation of a floating Ohmic contact (depicted in blue) connected to a single chiral, one-dimensional quantum Hall edge channel (illustrated by a dark gray arrow). Electrostatic modulation of the QH bulk regions with gates enables control over the trajectory of the edge state. The Ohmic contact resides at the interface between a turquoise QH bulk region and a light gray depleted region. The edge state originates from a large metallic contact (depicted in yellow) with applied voltage $V_{\rm in}$ and temperature $T_{\rm in}$. Fluctuations carried by the edge channel induce voltage fluctuations $\Delta V_{\rm oc}$ and temperature fluctuations $\Delta T_{\rm oc}$ in the Ohmic contact, subsequently leading to the emission of charge fluctuations $\delta I_{\rm out}$ and heat fluctuations $\delta J_{\rm out}$ in the outgoing channel

Table 1.1 Summary of relevant time scales and their corresponding energy scales. Charging energy estimate E_C is taken from [5, 50]. The level spacing estimate δE is from [50]. The estimate for the electron-phonon relaxation time $\tau_{\rm e-ph}$ is from [51]. The electron-electron scattering time $\tau_{\rm e-e}$ estimates are taken from [46–49]

Time scale	Typical magnitude	Energy scale
τ_C	10–100 ps	$E_C = 0.1$–1 K
$\tau_{\rm e-e}$	200 ps–100 ns	
τ_E	10–100 μs	$\delta E = 0.1$–1 μK
$\tau_{\rm e-ph}$	100 μs	at $T \approx 20$ mK

A second important timescale is the energy relaxation time,[2] denoted as τ_E, of the Ohmic contact, which is directly related to the finite heat capacitance of the OC. As discussed in Sect. 2.1, there is a natural separation of these timescales, with the energy relaxation time being much larger than the charge relaxation time ($\tau_C \ll \tau_E$). Although most previous works on the edge channel-OC system have focused on the transport of average charge and heat currents, as well as charge fluctuations, the heat current fluctuations emitted from the OC emitted from the OC are far less understood [9, 34–37, 39–44]. These fluctuations play a key role, for example, in the efficiency and limitations of nanoscale thermoelectric conversion [45]. A summary of the relevant time scales can be found in Table 1.1.

[2] Sometimes also referred to as Thouless time. It describes the lifetime of neutral excitations in the OC. For a more detailed discussion, see (2.8).

1.3 Energy Relaxation in QH Edges

The relaxation of energy in the QH edge states at a filling factor $\nu = 2$, was analyzed in [33]; see Fig. 1.2 for the schematic of the experimental setup. In these experiments, energy was introduced into a QH edge in the form of Joule heat, at a constant rate, through an upstream quantum point contact (QPC) with adjustable transmission. This QPC mixed two equilibrium Fermi distribution functions, each kept at different chemical potentials, creating a non-equilibrium "double step" distribution function. This distribution function then relaxes into a "hot" equilibrium Fermi distribution function, which could be probed by a downstream quantum dot. The goal was to study the relaxation process as a function of the distance between the injection and probe points. Depending on this distance, the measured distribution function exhibited characteristics of either the "double step" or the equilibrium Fermi distribution. The study [33] did not observe an energy transfer to the excitation of thermalized states and efficient energy redistribution between the two channels without particle exchanges. Interestingly, for long equilibration lengths $L \geq 10\,\mu\text{m}$, the temperature of the equilibrium distribution function reached a saturation point at a value 13% lower than anticipated for two interacting edge channels. This saturation of the effective temperature, regardless of the propagation length, suggests that no energy was lost to the bulk of the system or to any external mechanisms not accounted for by the experimental setup. This phenomenon is what we term the "*missing heat paradox*".

The authors concluded that the experiment's outcomes could be explained by the presence of additional degrees of freedom. However, for these observations to align with the experimental data, this additional mode would lead to a deviation from the established quantization principle of heat flux in one-dimensional systems.

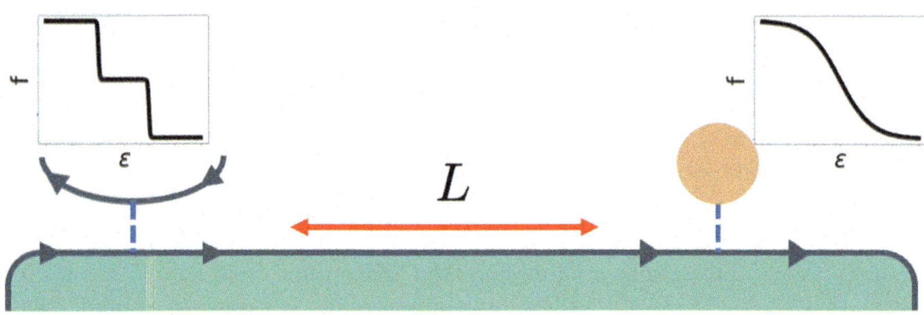

Fig. 1.2 Schematic of heat transport in a Quantum Hall (QH) edge state, as explored in the study by [33]. Energy is introduced via Joule heating through a Quantum Point Contact (QPC) on the left, generating a non-equilibrium "double step" distribution function. This function relaxes into a "hot" Fermi distribution, which is probed downstream by a quantum dot (depicted in orange). The QH bulk is shown in blue turquoise. The setup allows for adjusting the equilibration length L to investigate how the injected energy distribution relaxes over distance. The tunneling current $I_T \sim -df/d\varepsilon$, measured through a downstream quantum dot with a single active level (adjustable) provides a direct measure of the electronic distribution function's derivative in the channel. This measurement allows the calculation of the observed heat flux, as shown in Eq. (3.1)

Numerous experiments have shown the quantization of heat flux, indicating that the rate at which carriers can transport heat in a ballistic channel is a universal constant [53], known as the quantum heat flux:

$$J_q = \frac{\kappa_0}{2} T^2, \quad \kappa_0 = \frac{\pi}{6}, \tag{1.1}$$

where κ_0 is the thermal conductance quantum. However, the experimentally measured heat flux

$$J_{\text{meas}} = \int \frac{d\varepsilon}{2\pi} \, \varepsilon \, (f(\varepsilon) - \theta(-\epsilon)) \tag{1.2}$$

deviates from the quantum flux of heat $J_{\text{meas}} \neq J_q$, especially when $f(\varepsilon)$ is not a Fermi-Dirac distribution, suggesting a potential violation of this paradigm. Alternatively, incorporating dissipation into the system might solve this paradox by playing the role of additional degrees of freedom. Earlier theoretical investigations proposed that disspation in the low-energy theory of QH edge states could indeed alter the conventionally accepted notion of heat flux quantization [52].

In the following sections, we discuss the theoretical ideas derived from these experiments and outline where they can be found within this book.

1.4 Structure of the Book

This book is an extensive compilation of four projects that I have undertaken during my doctoral studies, each focusing on exploring different aspects of Ohmic contacts (OCs) in quantum hall edge systems. The essence and findings of these projects have been documented in individual papers, with this thesis aiming to present the information in a more detailed and accessible manner. This approach ensures that the thesis serves not only as a repository of condensed information for quick reference but also as an extensive resource for interested readers and future students. The objective is to improve their understanding of the subject matter, providing a clear explanation of the complex concepts and methodologies used in the articles.

In Chap. 2, we explore the concept of the OC as a fundamental component and theoretical model that plays a crucial role in contemporary quantum Hall experiments. Our discussion covers the theoretical model of the OC connected to quantum Hall (QH) edge states, followed by rederiving established results associated with the heat Coulomb blockade effect for both a single channel and multi-channel setup.

In Chap. 3, we introduce the transmission line (TL) model as a framework for understanding the phenomena of dissipation and potential non-local effects, such as diffusive currents, which can arise from disorder within QH edge states at a microscopic level. Our investigation begins with the "missing heat" paradox, where experimental observations of energy relaxation in QH edge states suggest an unaccounted loss of energy, possibly due to an unidentified mechanism. Although resolution of this paradox remains elusive, our analysis confirms that dissipation alone does not account for this discrepancy. We establish a connection between the condition of equilibrium heat flux within the edge state and a sum rule that ensures equilibrium is achieved. We explain why earlier theoretical approaches [52] were inadequate in accurately capturing the physics of dissipation in QH edge states.

In Chap. 4, we illustrate, in accordance with the second law of thermodynamics, how this aforementioned sum rule can be circumvented in an equilibrium drift-diffusion system. We explore how non-local couplings can facilitate the transfer of heat in other parts of the system, such that a QH edge state can carry an amount of heat that is greater than the expected equilibrium heat flux quantum. This phenomenon is attributed to additional correlations due to strong interactions.

In Chap. 5, we explore an OC that features edge states that loop back onto themselves. These self-looping configurations resemble the transmission line equipped with non-local couplings, discussed in Chap. 4. Beyond the expected contributions from free fermions, these loops carry additional correlations contributing to the heat flux due to the OC's self-interaction. The OC provides a framework for understanding the complex dynamics and correlations found in non-local TL systems. We demonstrate that although the edge state appears to be hotter than the equilibrium temperature of the device, extracting energy from it via a thermometer is impossible, in alignment with the principles of energy conservation. However, when these states are probed using a tunneling quantum point contact, they reveal unexpected behaviors. The correlated nature manifests itself through interaction-modified correlation functions, affecting electrical and thermal tunneling conductances and scaling dimensions. We reexamine the Wiedemann-Franz law, which relates electrical to thermal conductance. This ratio changes compared to the non-interacting setup in a surprisingly straightforward manner. The phenomena grouped as the "heat multiplier and fractionalizer" vary based on the device's setup, capable of producing states similar to fractional quantum Hall (FQH) states. The utility of this model in producing artificial FQH states emphasizes its significance and direct relevance to the field and provides an interesting foundation for further research.

In Chap. 6, we explore the heat Coulomb blockade's signature in heat noise power and temperature fluctuations of the OC due to its finite heat capacitance. Adopting a Langevin-like approach, we address thermal conductance and energy relaxation times, within the heat Coulomb blockade regime. Our analysis of the temperature-temperature correlation function reveals its universal equilibrium form, unaffected by the heat Coulomb blockade.

Furthermore, we discover that temperature fluctuations induce non-Gaussian Full Counting Statistics (FCS) of transmitted charge downstream of the device, underlining the deep interconnection between thermal fluctuations and quantum transport phenomena.

In the concluding Chap. 7, we present a collection of ideas and unfinished projects as both a launching point and an intellectual contribution to future research and the next generation of students. We include preliminary calculations on a nonchiral transmission line, discuss the fermionic distribution function in the presence of temperature fluctuations, and how to generalize the approach presented in Chap. 5 beyond the leading order in perturbation theory.

References

1. Klitzing KV, Dorda G, Pepper M (1980) New method for high-accuracy determination of the fine-structure constant based on quantized Hall resistance. Phys Rev Lett 45(6):494–497. https://doi.org/10.1103/PhysRevLett.45.494
2. Hasan MZ, Kane CL (2010) Colloquium: topological insulators. Rev Mod Phys 82:3045–3067. https://doi.org/10.1103/RevModPhys.82.3045
3. Duprez H, Pierre F, Sivre E, Aassime A, Parmentier FD, Cavanna A, Ouerghi A, Gennser U, Safi I, Mora C, Anthore A (2021) Dynamical Coulomb blockade under a temperature bias. Phys Rev Res 3:023122. https://doi.org/10.1103/PhysRevResearch.3.023122
4. Altimiras C, Parlavecchio O, Joyez P, Vion D, Roche P, Esteve D, Portier F (2014) Dynamical Coulomb blockade of shot noise. Phys Rev Lett 112:236803. https://doi.org/10.1103/PhysRevLett.112.236803
5. Sivre E, Anthore A, Parmentier FD, Cavanna A, Gennser U, Ouerghi A, Jin Y, Pierre F (2018) Heat Coulomb blockade of one ballistic channel. Nat Phys 14:145–148. https://doi.org/10.1038/nphys4280
6. Sivre E, Duprez H, Anthore A, Aassime A, Parmentier FD, Cavanna A, Ouerghi A, Gennser U, Pierre F (2019) Electronic heat flow and thermal shot noise in quantum circuits. Nat Commun 10:5638. https://doi.org/10.1038/s41467-019-13566-8
7. Idrisov EG, Levkivskyi IP, Sukhorukov EV (2018) Dephasing in a Mach-Zehnder interferometer by an Ohmic contact. Phys Rev Lett 121:026802. https://doi.org/10.1103/PhysRevLett.121.026802
8. Duprez H, Sivre E, Anthore A, Aassime A, Cavanna A, Gennser U, Pierre F (2019) Transmitting the quantum state of electrons across a metallic island with Coulomb interaction. Science 366:1243–1247. https://doi.org/10.1126/science.aaw7856
9. Idrisov EG, Levkivskyi IP, Sukhorukov EV (2022) Thermal drag effect in quantum Hall circuits. Phys Rev B 106:L121405. https://doi.org/10.1103/PhysRevB.106.L121405
10. Anthore A, Iftikhar Z, Boulat E, Parmentier FD, Cavanna A, Ouerghi A, Gennser U, Pierre F (2018) Circuit quantum simulation of a Tomonaga-Luttinger liquid with an impurity. Phys Rev X 8:031075. https://doi.org/10.1103/PhysRevX.8.031075
11. Jezouin S, Albert M, Parmentier FD, Anthore A, Gennser U, Cavanna A, Safi I, Pierre F (2013) Tomonaga-Luttinger physics in electronic quantum circuits. Nat Commun 4:1802. https://doi.org/10.1038/ncomms2810
12. Anthore A, Kennes DM, Boulat E, Andergassen S, Pierre F, Meden V (2020) Universality at work—the local sine-Gordon model, lattice fermions, and quantum circuits. Eur Phys J Spec Top 229:663–682. https://doi.org/10.1140/epjst/e2019-900117-5
13. Idrisov EG, Levkivskyi IP, Sukhorukov EV (2020) Quantum ammeter: measuring full counting statistics of electron currents at quantum timescales. Phys Rev B 101:245426. https://doi.org/10.1103/PhysRevB.101.245426
14. Morel T, Lee J-YM, Sim H-S, Mora C (2022) Fractionalization and anyonic statistics in the integer quantum Hall collider. Phys Rev B 105:075433. https://doi.org/10.1103/PhysRevB.105.075433

15. Iftikhar Z, Jezouin S, Anthore A, Gennser U, Parmentier FD, Cavanna A, Pierre F (2015) Two-channel Kondo effect and renormalization flow with macroscopic quantum charge states. Nature 526:233–236. https://doi.org/10.1038/nature15384
16. Iftikhar Z, Anthore A, Mitchell AK, Parmentier FD, Gennser U, Ouerghi A, Cavanna A, Mora C, Simon P, Pierre F (2018) Tunable quantum criticality and super-ballistic transport in a "charge" Kondo circuit. Science 360:1315–1320. https://doi.org/10.1126/science.aan5592
17. Kamata H, Kumada N, Hashisaka M, Muraki K, Fujisawa T (2014) Fractionalized wave packets from an artificial Tomonaga-Luttinger liquid. Nat Nanotechnol 9:177–181. https://doi.org/10.1038/nnano.2013.312
18. Berg E, Oreg Y, Kim E-A, von Oppen F (2009) Fractional charges on an integer quantum Hall edge. Phys Rev Lett 102:236402. https://doi.org/10.1103/PhysRevLett.102.236402
19. Béri B (2017) Exact nonequilibrium transport in the topological Kondo effect. Phys Rev Lett 119:027701. https://doi.org/10.1103/PhysRevLett.119.027701
20. Landau LA, Cornfeld E, Sela E (2018) Charge fractionalization in the two-channel Kondo effect. Phys Rev Lett 120:186801. https://doi.org/10.1103/PhysRevLett.120.186801
21. Inoue H, Grivnin A, Ofek N, Neder I, Heiblum M, Umansky V, Mahalu D (2014) Charge fractionalization in the integer quantum Hall effect. Phys Rev Lett 112:166801. https://doi.org/10.1103/PhysRevLett.112.166801
22. Neder I, Marquardt F, Heiblum M, Mahalu D, Umansky V (2007) Controlled dephasing of electrons by non-Gaussian shot noise. Nat Phys 3:534–537. https://doi.org/10.1038/nphys627
23. Roulleau P, Portier F, Roche P, Cavanna A, Faini G, Gennser U, Mailly D (2009) Tuning decoherence with a voltage probe. Phys Rev Lett 102:236802. https://doi.org/10.1103/PhysRevLett.102.236802
24. Roulleau P, Portier F, Roche P, Cavanna A, Faini G, Gennser U, Mailly D (2008) Direct measurement of the coherence length of edge states in the integer quantum Hall regime. Phys Rev Lett 100:126802. https://doi.org/10.1103/PhysRevLett.100.126802
25. Huynh P-A, Portier F, le Sueur H, Faini G, Gennser U, Mailly D, Pierre F, Wegscheider W, Roche P (2012) Quantum coherence engineering in the integer quantum Hall regime. Phys Rev Lett 108:256802. https://doi.org/10.1103/PhysRevLett.108.256802
26. Bieri E, Weiss M, Göktas O, Hauser M, Schönenberger C, Oberholzer S (2009) Finite-bias visibility dependence in an electronic Mach-Zehnder interferometer. Phys Rev B 79:245324. https://doi.org/10.1103/PhysRevB.79.245324
27. Litvin LV, Helzel A, Tranitz H-P, Wegscheider W, Strunk C (2008) Edge-channel interference controlled by Landau level filling. Phys Rev B 78:075303. https://doi.org/10.1103/PhysRevB.78.075303
28. Litvin LV, Tranitz H-P, Wegscheider W, Strunk C (2007) Decoherence and single electron charging in an electronic Mach-Zehnder interferometer. Phys Rev B 75:033315. https://doi.org/10.1103/PhysRevB.75.033315
29. Helzel A, Litvin LV, Levkivskyi IP, Sukhorukov EV, Wegscheider W, Strunk C (2015) Counting statistics and dephasing transition in an electronic Mach-Zehnder interferometer. Phys Rev B 91:245419. https://doi.org/10.1103/PhysRevB.91.245419
30. Neder I, Heiblum M, Levinson Y, Mahalu D, Umansky V (2006) Unexpected behavior in a two-path electron interferometer. Phys Rev Lett 96:016804. https://doi.org/10.1103/PhysRevLett.96.016804
31. Altimiras C, le Sueur H, Gennser U, Cavanna A, Mailly D, Pierre F (2010) Tuning energy relaxation along quantum Hall channels. Phys Rev Lett 105:226804. https://doi.org/10.1103/PhysRevLett.105.226804
32. Altimiras C, le Sueur H, Gennser U, Cavanna A, Mailly D, Pierre F (2010) Non-equilibrium edge-channel spectroscopy in the integer quantum Hall regime. Nat Phys 6:34–39. https://doi.org/10.1038/nphys1429
33. le Sueur H, Altimiras C, Gennser U, Cavanna A, Mailly D, Pierre F (2010) Energy relaxation in the integer quantum Hall regime. Phys Rev Lett 105:056803. https://doi.org/10.1103/PhysRevLett.105.056803

34. Pekola JP, Karimi B (2021) Colloquium: quantum heat transport in condensed matter systems. Rev Mod Phys 93:041001. https://doi.org/10.1103/RevModPhys.93.041001
35. Krive IV, Bogachek EN, Scherbakov AG, Landman U (2001) Heat current fluctuations in quantum wires. Phys Rev B 64:233304. https://doi.org/10.1103/PhysRevB.64.233304
36. Sergi D (2011) Energy transport and fluctuations in small conductors. Phys Rev B 83:033401. https://doi.org/10.1103/PhysRevB.83.033401
37. Battista F, Haupt F, Splettstoesser J (2014) Energy and power fluctuations in ac-driven coherent conductors. Phys Rev B 90:085418. https://doi.org/10.1103/PhysRevB.90.085418
38. Crépieux A, Michelini F (2014) Mixed, charge and heat noises in thermoelectric nanosystems. J Phys: Condens Matter 27:015302. https://doi.org/10.1088/0953-8984/27/1/015302
39. Ebisu H, Schiller N, Oreg Y (2022) Fluctuations in heat current and scaling dimension. Phys Rev Lett 128:215901. https://doi.org/10.1103/PhysRevLett.128.215901
40. Crépieux A (2021) Electronic heat current fluctuations in a quantum dot. Phys Rev B 103:045427. https://doi.org/10.1103/PhysRevB.103.045427
41. Averin DV, Pekola JP (2010) Violation of the fluctuation-dissipation theorem in time-dependent mesoscopic heat transport. Phys Rev Lett 104:220601. https://doi.org/10.1103/PhysRevLett.104.220601
42. Dashti N, Misiorny M, Samuelsson P, Splettstoesser J (2018) Probing charge- and heat-current noise by frequency-dependent fluctuations in temperature and potential. Phys Rev Appl 10:024007. https://doi.org/10.1103/PhysRevApplied.10.024007
43. Eriksson J, Acciai M, Tesser L, Splettstoesser J (2021) General bounds on electronic shot noise in the absence of currents. Phys Rev Lett 127:136801. https://doi.org/10.1103/PhysRevLett.127.136801
44. Tesser L, Acciai M, Spånslätt C, Monsel J, Splettstoesser J (2023) Charge, spin, and heat shot noises in the absence of average currents: conditions on bounds at zero and finite frequencies. Phys Rev B 107:075409. https://doi.org/10.1103/PhysRevB.107.075409
45. Benenti G, Casati G, Saito K, Whitney RS (2017) Fundamental aspects of steady-state conversion of heat to work at the nanoscale. Phys Rep 694:1–124. https://doi.org/10.1016/j.physrep.2017.05.008
46. Katine JA, Berry MJ, Westervelt RM, Gossard AC (1998) Determination of the electronic phase coherence time in one-dimensional channels. Phys Rev B 57(3):1698–1702. https://doi.org/10.1103/PhysRevB.57.1698
47. Pierre F, Gougam AB, Anthore A, Pothier H, Esteve D, Birge NO (2003) Dephasing of electrons in mesoscopic metal wires. Phys Rev B 68(8):085413. https://doi.org/10.1103/PhysRevB.68.085413
48. Song JCW, Tielrooij KJ, Koppens FHL, Levitov LS (2013) Photoexcited carrier dynamics and impact-excitation cascade in graphene. Phys Rev B 87(15):155429. https://doi.org/10.1103/PhysRevB.87.155429
49. Song JCW, Levitov LS (2015) Energy flows in graphene: hot carrier dynamics and cooling. J Phys: Condens Matter 27(16):164201. https://doi.org/10.1088/0953-8984/27/16/164201
50. Jezouin S, Parmentier FD, Anthore A, Gennser U, Cavanna A, Jin Y, Pierre F (2013) Quantum limit of heat flow across a single electronic channel. Science 342:601–604. https://doi.org/10.1126/science.1241912
51. Gasparinetti S, Viisanen KL, Saira O-P, Faivre T, Arzeo M, Meschke M, Pekola JP (2015) Fast electron thermometry for ultrasensitive calorimetric detection. Phys Rev Appl 3(1):014007. https://doi.org/10.1103/PhysRevApplied.3.014007
52. Goremykina A, Borin A, Sukhorukov E (2019) Heat current in a dissipative quantum Hall edge. arXiv:1908.01213
53. Pendry JB (1983) Quantum limits to the flow of information and entropy. J Phys A: Math Gen 16:2161–2171. https://doi.org/10.1088/0305-4470/16/10/012

Chapter 2
The Ohmic Contact

Abstract This chapter explores the physics of Ohmic contacts (OCs) within the quantum Hall regime, focusing on their role in charge and heat transport in mesoscopic systems. The OC is modeled as a floating metallic reservoir interacting with quantum Hall edge channels, leading to significant charge and energy fluctuations. By analyzing the system's charge and heat dynamics, we derive the specific heat, energy relaxation times, and key noise properties of the OC. The framework incorporates bosonic field theory and Langevin equations to describe charge conservation and thermal fluctuations, ultimately leading to an explicit relation between charge and heat currents. We further investigate the heat Coulomb blockade effect, where charge quantization and finite capacitance limit heat equilibration. Extending the analysis to multiple edge channels, we demonstrate how the OC modifies energy transport and affects heat conductance quantization. These insights are crucial for understanding the mechanisms of dissipation and energy distribution in low-dimensional electron systems, with implications for quantum thermodynamics and electronic cooling in mesoscopic devices.

2.1 Setup and Time-Scales

In the system depicted in Fig. 1.1, we examine a floating Ohmic contact (OC) connected to a single quantum Hall edge channel originating from another large metallic contact, viewed as an ideal, large Fermi reservoir of electrons. The reservoir voltage and temperature are denoted as V_{in} and T_{in}, respectively, both with negligible fluctuations. For the purpose of this work, unless otherwise specified, we assume $V_{\text{in}} = 0$. Given the chiral nature of the connected edge channel, it features one incoming and one outgoing segment relative to the OC. The charge and heat currents transported by the incoming channel can cause significant fluctuations in the OC voltage $V_{\text{oc}}(t) \equiv V_{\text{oc}} + \Delta V_{\text{oc}}(t)$ and OC temperature $T_{\text{oc}}(t) \equiv T_{\text{oc}} + \Delta T_{\text{oc}}(t)$, attributed to the finite charge and energy relaxation times of the OC, τ_C and τ_E, respectively.

F. Stäbler, *Dissipation, Interactions, and Heat Transport in Quantum Hall Devices*, Springer Theses, https://doi.org/10.1007/978-3-031-90671-8_2

2.1.1 Specific Heat and Energy Relaxation Time of an Ohmic Contact

We calculate the heat capacitance, $C_E[T_{\rm oc}]$ and the energy relaxation time, τ_E, assuming that the OC resembles a free three-dimensional (3D) Fermi gas with volume L^3 and at temperature T. We also assume the OC exhibits quantum, not classical,[1] behavior, especially at low temperatures where $T_{\rm oc} \ll E_F$, with the Fermi energy $E_F = k_F^2/2m$, k_F being the Fermi momentum, and m the electron mass.

For a Fermi gas, the increase in internal energy, $\Delta U(T_{\rm oc})$, due to a finite temperature $T_{\rm oc}$ is given by

$$\Delta U(T_{\rm oc}) = \int_0^{\infty} dE\, E\mathcal{D}(E) f(E) - \int_0^{E_F} dE\, E\mathcal{D}(E), \tag{2.1}$$

where $f(E) = (e^{(E-\mu)/T_{\rm oc}} + 1)^{-1}$ is the Fermi-Dirac distribution, μ is the electrochemical potential, and $\mathcal{D}(E)$ is the density of states. For a 3D gas, $\mathcal{D}(E) \sim \sqrt{E}$. Since the number of electrons, $\mathcal{N} = L^3 k_F^3/(3\pi^2)$, is conserved at both zero and finite temperature, we find the identity

$$\int_0^{E_F} dE\, \mathcal{D}(E)(f(E) - 1) + \int_{E_F}^{\infty} dE\, \mathcal{D}(E) f(E) = 0. \tag{2.2}$$

Then, by differentiating Eq. (2.1) with respect to temperature and utilizing Eq. (2.2), we find the specific heat $C_E[T_{\rm oc}] \equiv d\Delta U(T_{\rm oc})/dT_{\rm oc}$

$$C_E[T_{\rm oc}] = \int_{E_F}^{\infty} dE(E - E_F)\mathcal{D}(E)\frac{\partial f(E)}{\partial T_{\rm oc}} = \frac{\pi^2}{2}\frac{T_{\rm oc}}{T_F}\mathcal{N}, \tag{2.3}$$

assuming the density of states near the Fermi energy is given by, $\mathcal{D}(E) \approx \mathcal{D}(E_F) \equiv (2m)^{3/2}/(2\pi^2)\sqrt{E_F}$, and noting that in our unit convention ($e = \hbar = k_B = 1$), the Fermi temperature $T_F = E_F$.

As our final step, we compare $C_E[T]$ with the level spacing, $\delta E_{\rm lvl}$, which for a D-dimensional Fermi gas of size L^D is given by the inverse density of states at the Fermi level $1/\mathcal{D}(E_F)$. Hence,

$$\delta E_{\rm lvl} \equiv \frac{1}{\mathcal{D}(E_F)} = \frac{2E_F}{3\mathcal{N}} = \frac{1}{(3\pi^2\mathcal{N})^{\frac{1}{3}}}\frac{\pi^2}{mL^2}, \tag{2.4}$$

which has a characteristic energy scale $\pi^2/(mL^2)$ and decreases with increasing electron number $\mathcal{N}$. Combining this result with Eq. (2.3), gives the scaling $C_E[T] \sim$

[1] We neglect the temperature dependence of the Fermi energy, which gives a subleading correction to the specific heat for $T_{\rm oc} \ll E_F$.

$T\delta E_{\mathrm{lvl}}^{-1}$ and the energy relaxation time thus increases linearly with the inverse level spacing:

$$\tau_E \equiv \frac{C_E[T]}{\kappa_0 T} \sim \delta E_{\mathrm{lvl}}^{-1}, \tag{2.5}$$

where $\kappa_0 = \pi/6$ is the thermal conductance quantum. Note that for a 2D Fermi gas the level spacing (2.4) is given by the same formula

$$\delta E_{\mathrm{lvl},2D} \equiv \frac{1}{\mathcal{D}_{2D}(E_F)} = \frac{E_F}{\mathcal{N}_{2D}} = \frac{1}{\pi}\frac{\pi^2}{mL^2}, \tag{2.6}$$

with $\mathcal{D}_{2D}(E_F) = E_F/\mathcal{N}_{2D}$ and $\mathcal{N}_{2D} = k_F^2 L^2/2\pi$. Note that the 2D level spacing is independent of the electron number $\mathcal{N}_{2D}$. In the 1D case we find

$$\delta E_{\mathrm{lvl},1D} \equiv \frac{1}{\mathcal{D}_{1D}(E_F)} = \frac{E_F}{2\mathcal{N}_{1D}} = \frac{\mathcal{N}_{1D}}{4}\frac{\pi^2}{mL^2}, \tag{2.7}$$

with $\mathcal{D}_{1D}(E_F) = E_F/(2\mathcal{N}_{1D})$ and $\mathcal{N}_{1D} = k_F L/\pi$. Note that the 1D level spacing increases linearly with the electron number $\mathcal{N}_{1D}$.

Typically, the energy required to add an electron to the OC, the charging energy E_C, scales linearly with the inverse size of the OC L^{-1}, through its capacitance C: Approximately, we have $E_C \sim e^2/C \sim e^2/(\varepsilon L) \approx 0.1 - 1\mathrm{meV} \approx 1 - 10\mathrm{K}$, for a typical semiconducting dielectric constant $\varepsilon = 10$. In contrast, the energy relaxation time, τ_E, is proportional to the inverse level spacing of the OC, $\delta E_{\mathrm{lvl}}^{-1}$, when the OC is viewed as a 3D metallic island: $\tau_E = C_E[T_{\mathrm{oc}}(t)]/(\kappa_0 T_{\mathrm{oc}}) \sim \delta E_{\mathrm{lvl}}^{-1}$. The level spacing of a typical micrometer-sized OC is estimated to be $\delta E_{\mathrm{lvl}} \approx 0.1 - 1\mu K$. To achieve a higher $\delta E_{\mathrm{lvl}} \sim 1\mathrm{K}$ in a 3D metallic island with Fermi wavelength $\lambda_F \sim 1\mathrm{nm}$, the typical quantum dot (QD) size required would be $L \sim 10\mathrm{nm}$. For a 2D QD, $L \sim 100\mathrm{nm}$ suffices. The above estimates agree with the charging energy $E_C \approx 0.3K$ and level spacing $\delta E_{\mathrm{lvl}} \approx 0.2\mu K$ that were observed in experiments on micrometer-sized OCs [2]. This implies that in this work, we assume a separation of time scales

$$\tau_C \ll \tau_{\mathrm{e-e}} \ll \tau_E \ll \tau_{\mathrm{e-ph}}, \tag{2.8}$$

which is motivated by the above considerations and supported by the estimates in Table 1.1. The first inequality allows the description of charge fluctuations in the quasi-particle picture as a collective mode. It implies that $\tau_{\mathrm{e-e}}^{-1} \sim T^2/E_F \rightarrow T \ll E_C$, which is the interesting (heat) Coulomb blockade regime, we focus on in this thesis.[2] The second inequality assumes a short electron-electron relaxation time $\tau_{\mathrm{e-e}}$ compared to the energy relaxation or dwell time τ_E of the electrons inside of the reservoir and implies that the state of the OC can be characterized by an equilibrium

[2] The non heat Coulomb blockade regime is given by $E_C \ll T \ll E_C E_F/T$.

Fermi distribution function. Meanwhile, the final inequality assumes a large electron-phonon relaxation time $\tau_{\text{e-ph}}$, such that no energy is dissipated to phonons. The heat flux carried by phonons typically scales as T^α, with $\alpha > 2$, which implies that it can be neglected for temperatures that are low enough. The latter condition could be relaxed by directly modelling this energy loss. In most cases, if not specified otherwise, we do not consider temperature fluctuations and set $\tau_E \to \infty$.

2.2 Bosonic Model of the Ohmic Contact and Derivation of the Heat-Charge Current Relation

Consider the scenario where the temperature of the OC remains constant, which implies $\tau_E \to \infty$. For a single edge state interacting with the OC, we express the bosonic Hamiltonian of the edge-OC system, as shown in the works of [7–9] as

$$\mathcal{H} = \frac{v_F}{4\pi}\sum_\sigma \int_{-\infty}^{\infty} dx\, (\partial_x \phi_\sigma(x,t))^2 + \frac{Q^2(t)}{2C}, \tag{2.9}$$

where $\sigma = \pm$ labels the outgoing and incoming edge channels, C is the OC capacitance, and $Q(t) = \sum_{\sigma'} \int_{-\infty}^{0} dx\, \rho_{\sigma'}(x,t) e^{\varepsilon x/v_F}$ is the OC charge operator, see Fig. 2.1 for a schematic setup. The exponentially decaying term arises because of the finite lifetime of excitations $\varepsilon > 0$ inside the OC. Here, the charge is expressed as integrated charge densities $\rho_\sigma(x,t) \equiv \frac{1}{2\pi}\partial_x\phi_\sigma(x,t)$ within the OC region $x \in (-\infty, 0]$. Current densities may be expressed in similar terms as $I_\sigma(x,t) \equiv -\frac{1}{2\pi}\partial_t\phi_\sigma(x,t)$. These boson fields satisfy the following canonical commutation relations:

$$[\phi_\sigma(x,t), \phi_{\sigma'}(y,t)] = i\pi\sigma\,\text{sgn}(x-y), \tag{2.10}$$

where $\text{sgn}(x-y)$ denotes the sign function. Importantly, the Hamiltonian (2.9) is quadratic in the bosonic fields, and the effect of charge capacitance can be taken into account exactly. The equations of motion for the bosons are obtained from the Heisenberg equation $\partial_t\phi_\sigma(x,t) = -i\left[\phi_\sigma(x,t), \mathcal{H}\right]$, and yield after applying the commutation relations (2.10)

$$\sigma\partial_t\phi_\sigma(x,t) + v_F\partial_x\phi_\sigma(x,t) = -\frac{Q(t)}{C} e^{\varepsilon x/v_F}\theta(-x), \tag{2.11}$$

where $\theta(x)$ is the Heaviside step function. These equations can be solved by applying a Fourier transform from time to frequency domain and integrating the resulting differential equation with respect to x.

The general solution for $x \le 0$ is

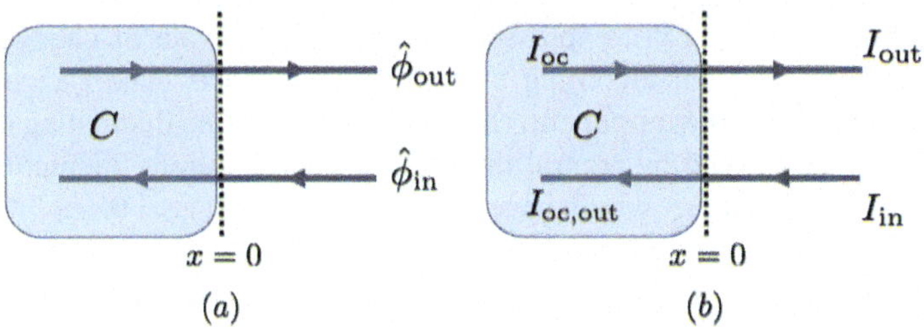

Fig. 2.1 Schematic picture of an Ohmic contact. Panel **a** indicates the chirality of the fields ϕ_σ. Panel **b** indicates the outgoing and incoming currents of the OC

$$\phi_\sigma(x,\omega) = \phi_\sigma(\omega)e^{i\sigma\omega x/v_F} + \frac{\sigma}{i\omega\tau_C}\sum_{\sigma'}\phi_{\sigma'}(\omega)e^{\varepsilon x/v_F}, \tag{2.12}$$

where $\phi_\sigma(\omega)$ are the constants of integration. Introducing the following boundary conditions for the currents $\partial_t\phi_+(-\infty,t) = -2\pi I_{oc}(t)$, $\partial_t\phi_-(0,t) = 2\pi I_{in}(t)$, $\partial_t\phi_+(0,t) = -2\pi I_{out}(t)$ and $\partial_t\phi_-(-\infty,t) = 2\pi I_{oc,out}(t)$, allows to reformulate the solution of the equation of motion (2.12) in terms of a scattering problem, i.e. expressing the outgoing currents in terms of the incoming (boundary) currents

$$I_{out}(\omega) = \sum_{p=\text{in,oc}} \mathcal{T}_p(\omega)I_p(\omega), \tag{2.13a}$$

$$\mathcal{T}_{in}(\omega) = 1 - \mathcal{T}_{oc}(\omega) = (1 - i\omega\tau_C)^{-1}. \tag{2.13b}$$

Figure 2.1, shows the labeling of the boundary terms. The frequency-dependent coefficients $\mathcal{T}_p(\omega)$, with $p \in \{\text{in, oc}\}$, can be viewed as scattering amplitudes of bosonic density fluctuations carried along the chiral edge channel and by excitations inside the OC.

2.2.1 *Langevin Equations*

The most striking feature of the solution of the equation of motion (2.13a) is that it can be obtained in an alternative way from circuit theory. This motivates an analysis of the same system (Fig. 2.1) by considering charge conservation, which yields a relationship between the temporal OC charge $Q(t)$ and the incoming ($I_{in}(t)$) and outgoing ($I_{out}(t)$) edge charge currents as follows:

$$\frac{dQ(t)}{dt} = I_{in}(t) - I_{out}(t), \tag{2.14a}$$

$$I_{out}(t) = \frac{Q(t)}{\tau_C} + I_{oc}(t). \tag{2.14b}$$

Here, Kirchoff's law (2.14a), expresses the OC charge's rate of change in terms of the difference between incoming and outgoing currents. The Langevin equation (2.14b), describes the output current as influenced by the fluctuating OC potential $V_{\mathrm{oc}}(t) \equiv Q(t)/C$ and by neutral thermally induced current fluctuations $I_{\mathrm{oc}}(t)$, called the Langevin source, which have zero average: $\langle I_{\mathrm{oc}}(t)\rangle = 0$.

2.2.2 *Definition of the Heat Current*

By inspecting the first term of the edge Hamiltonian (2.9), we obtain the heat current operator for a single channel, J, as follows. In 1D, this operator must have the form of a velocity multiplied by the energy density, for which the only choice is

$$J(x,t) = \frac{v_F^2}{4\pi}\left(\partial_x \phi(x,t)\right)^2 . \tag{2.15}$$

Identifying in this expression the current operator $I(x,t) = v_F \rho(x,t) = e v_F \partial_x \phi(x,t)/(2\pi)$ and the resistance quantum $R_q = 2\pi\hbar/e^2$, leads directly to the heat-charge current relation

$$J(x,t) = \frac{R_q}{2} I^2(x,t). \tag{2.16}$$

This can be further motivated by the following gedankenexperiment. Consider a section of edge state $a < x < b$. We define a continuity equation for heat in the usual way by $\partial_x J = -\partial_t \langle \mathcal{H} \rangle$, which describes the gradient of net heat flux as the rate of change of energy of the edge segment. In a steady state, the rate of change of energy is zero, i.e., the incoming and outgoing heat fluxes are equal, effectively demonstrated by the expression for the heat flux (2.16) presented earlier, verifiable by explicitly computing $\partial_t \langle \mathcal{H} \rangle$.

$$\langle J \rangle_{\mathrm{net}} = -\int_a^b dx\, \partial_t \langle \mathcal{H} \rangle = \left[\frac{R_q}{2}\left\langle I^2(x,t)\right\rangle\right]\Bigg|_a^b = 0. \tag{2.17}$$

Since the net heat current consists of incoming minus outgoing heat flux, one can identify the correct heat current operator per cross section (2.16).

The heat-charge current relation (2.16) allows us to compute e.g. the average heat current $\langle J(x,t)\rangle = \frac{R_q}{2}\left\langle I^2(x,t)\right\rangle$ in terms of the current-current correlation function.[3] Although making assumptions about the correlation function of the outgoing charge current is hard and not intuitive, we can rely on the scattering formalism presented in Eqs. 2.13a and 2.13b. The outgoing current I_{out} is expressed in terms of boundary

[3] To streamline our notation throughout, we will generally omit the bracket notation for the average heat current, identifying $\langle J(x,t)\rangle \equiv J$ for simplicity.

currents I_{in} and I_{oc} emanating from the source contacts or deep inside the OC and can be viewed as equilibrium currents with a well-defined noise power spectral density $S_p(\omega)$ characterized by a local equilibrium temperature T_p, for $p \in \{\text{in}, \text{oc}\}$.

2.3 Output Charge Current and Charge Current Noise Power

Starting with a Fourier transform to solve the charge dynamics Eqs. (2.14a) and (2.14b) for the output charge current $I_{\text{out}}(\omega)$ in frequency space yields a solution as a linear combination of input and OC currents

$$I_{\text{out}}(\omega) = \sum_{p=\text{in,oc}} \mathcal{T}_p(\omega) I_p(\omega), \tag{2.18a}$$

$$\mathcal{T}_{\text{in}}(\omega) = 1 - \mathcal{T}_{\text{oc}}(\omega) = (1 - i\omega\tau_C)^{-1}. \tag{2.18b}$$

We see that current conservation leads to equal average input and output charge currents at zero frequency.

$$I_{\text{out}}(\omega = 0) = I_{\text{in}}(\omega = 0). \tag{2.19}$$

Next, we define the output charge noise power (non-symmetrized) $S_{\text{out}}(\omega)$ defined through the two-point correlation function of output fluctuations $\delta I_{\text{out}}(\omega) \equiv I_{\text{out}}(\omega) - \langle I_{\text{out}}(\omega) \rangle$, laying the groundwork for understanding the characteristics of the noise.

$$\langle \delta I_{\text{out}}(\omega) \delta I_{\text{out}}(\omega') \rangle \equiv 2\pi \delta(\omega + \omega') S_{\text{out}}(\omega). \tag{2.20}$$

The OC is described by a local equilibrium state, so that all averages $\langle \cdots \rangle$ are taken with respect to the equilibrium state of the input and OC (the input channel emanates from an equilibrium Fermi reservoir). By combining the output current solution (2.13a) and the noise power definition (2.20), we find

$$S_{\text{out}}(\omega) = \sum_{p=\text{in,oc}} |\mathcal{T}_p(\omega)|^2 S_p(\omega) = \frac{S_{\text{in}}(\omega) + S_{\text{oc}}(\omega)\omega^2\tau_C^2}{1 + \omega^2\tau_C^2}, \tag{2.21}$$

where $S_p(\omega)$ is the charge noise power of subsystem p, given by the equilibrium form

$$S_p(\omega) = \frac{\omega R_q^{-1}}{1 - e^{-\omega/T_p}}. \tag{2.22}$$

Here, T_p is the input channel temperature for $p = \text{in}$ and is the OC temperature for $p = \text{oc}$. It follows from the scattering amplitude definitions (2.13b) that

$$\sum_{p=\text{in},\text{oc}} |\mathcal{T}_p(\omega)|^2 = 1, \tag{2.23}$$

which reflects charge conservation. Throughout this book, we refer to this as the *sum rule*. We further note that $S_{\text{out}}(\omega)$ in Eq. (2.21) is in equilibrium form (2.22) if and only if $T_{\text{in}} = T_{\text{oc}}$ or in the asymptotic limits $\tau_C \to 0, \infty$. For different temperatures $T_{\text{in}} \neq T_{\text{oc}}$, the output noise power $S_{\text{out}}(\omega)$ is a weighted sum given by Eq. 2.21 indicating that the OC acts as a band-pass filter for the chiral edge channel.

2.3.1 Effective Output Temperature and Output Heat Current

The output noise power (2.21) can be used to define an effective temperature for the output channel, which is not in thermal equilibrium. It is important to note that "effective" in this context means that this temperature is more of a parameter that characterizes certain properties of the output noise power, rather than an actual temperature as understood in equilibrium thermodynamics. One might consider using the Johnson-Nyquist relation [10, 11] to define a Johnson-Nyquist temperature for the output, represented as $T_{\text{out}}^{\text{JN}} \equiv R_q S_{\text{out}}(0)$. However, this turns out to be not particularly useful for a setup involving only a single channel, since

$$T_{\text{out}}^{\text{JN}} = R_q S_{\text{out}}(0) = R_q \sum_{p=\text{in},\text{oc}} \mathcal{T}_p(0) S_p(0) = R_q (R_q^{-1} T_{\text{in}} + 0) = T_{\text{in}}. \tag{2.24}$$

This is because the Johnson-Nyquist temperature $T_{\text{out}}^{\text{JN}}$ ends up being equal to the input temperature T_{in}, as charge conservation dictates that output fluctuations at zero frequency $\omega = 0$ must match input fluctuations at the same frequency, see Eq. (2.19), which are characterized by the input temperature T_{in}. This is due to the requirement of the OC being floating and connected to only a single chiral edge channel. Hence, over long timescales, charge fluctuations do not "see" the effects of the OC capacitance. This explains why phenomena such as the dynamical (heat) Coulomb blockade [1, 2] are termed "dynamical", reflecting their dependence on frequency.

The limitations associated with the Johnson-Nyquist temperature, as outlined in Eq. (2.24), require an alternative and more useful definition of the output temperature. To achieve this, we start with the equilibrium relationship between the one-dimensional heat current J_{eq} and the temperature T:

$$J_{\text{eq}} = \int_0^\infty \frac{d\omega}{2\pi} \omega n_B(\omega) = \frac{R_q}{4\pi} \int_{-\infty}^\infty d\omega \left[S(\omega) - \frac{\omega\theta(\omega)}{R_q} \right] = \frac{\kappa_0}{2} T^2. \tag{2.25}$$

This definition is intimately related with the earlier definition of the heat current operator for a ballistic chiral edge state (2.16), since, in equilibrium, the noise power

$S(\omega)$ in Eq. (2.25) can be obtained from the Fourier transform of the current-current correlation function in (2.16).

In this derivation, we initially describe the edge channel heat current as the thermal energy carried by bosonic fluctuations, which is characterized in terms of the Bose-Einstein distribution $n_B(\omega) \equiv (\exp(\omega/T) - 1)^{-1}$. This definition is equivalent to the fermionic/measured heat flux (1.2), see Appendix D.1. The second equality relates the Bose-Einstein distribution to the equilibrium noise power as detailed in Eq. (2.22). The expression $\omega\theta(\omega)/R_q$, which involves the step function $\theta(\omega)$, is interpreted as subtracting vacuum fluctuations, which is the zero temperature contribution. The equilibrium relation expressed in Eq. (2.25) is valid for any Abelian 1D chiral channel,[4] regardless of its charge or whether it is fractional or not, since the factor R_q, and consequently, the central charge of the edge state is fully canceled, cf. Eq. (2.22).

We now propose defining an effective temperature through an expression analogous to Eq. (2.25), but for a *generic noise power*—that is, one not necessarily in equilibrium form. For the setup depicted in Fig. 1.1, we express the output heat current as the integral over the output charge current noise power:

$$J_{\text{out}} = \frac{R_q}{4\pi} \int_{-\infty}^{\infty} d\omega \left[S_{\text{out}}(\omega) - \frac{\omega\theta(\omega)}{R_q} \right] \equiv \frac{\kappa_0}{2} T_{\text{out}}^2. \tag{2.26}$$

The sum rule (2.23), ensures the convergence of the integral in Eq. (2.26). To explicitly compute the effective output temperature[5] T_{out}, we insert the output noise power (2.21) into Eq. (2.26) and incorporate the scattering amplitudes (2.13b) along with the equilibrium noise powers (2.22). After a series of algebraic manipulations, we derive the following.

$$T_{\text{out}}^2 = T_{\text{oc}}^2 + \frac{6}{\tau_C^2 \pi^2} \left[\mathcal{F}\left(\frac{1}{\tau_C T_{\text{in}}} \right) - \mathcal{F}\left(\frac{1}{\tau_C T_{\text{oc}}} \right) \right], \tag{2.27}$$

which is expressed in terms of the dimensionless integral function,

$$\mathcal{F}(a) \equiv \int_0^{\infty} \frac{z dz}{z^2 + a^2} \frac{1}{e^z - 1} = \frac{1}{2} \left[\ln\left(\frac{a}{2\pi} \right) - \frac{\pi}{a} - \psi\left(\frac{a}{2\pi} \right) \right]. \tag{2.28}$$

[4] Notably, for elementary fractionally charged edge states, one would adjust $R_q \to R_q/\nu$, where ν denotes the filling factor. This modification does not affect the expression of the average heat flux. For more complex edge reconstructions, the heat flux is influenced by the total number of channels in the reconstruction, denoted ν_Q.

[5] Note that T_{out} serves not only as a parameter for the outgoing heat current but also can be interpreted as the local equilibrium temperature of the edge state after its relaxation into an equilibrium state, due to the presence of disorder [5, 12]. This interpretation arises because the edge state remains isolated, "floating", ensuring the conservation of energy except for its dissipation to a bulk phonon bath, a phenomenon not modeled in our current framework. Given that these relaxation mechanisms are not well understood, being contingent upon many factors and specific experimental configurations, we do not dwell on this fact further.

Here, $\psi(z)$ denotes the logarithmic derivative of the gamma function. Two useful limits of $\mathcal{F}(a)$ are

$$\mathcal{F}(a \gg 1) \approx \frac{\pi^2}{6a^2} - \frac{\pi^4}{15a^4}, \tag{2.29a}$$

$$\mathcal{F}(a \ll 1) \approx \frac{\pi}{2a}. \tag{2.29b}$$

Equation (2.27) shows that T_{out} is directly influenced by the interaction between the RC time τ_C and the thermal time scales determined by the input and OC temperatures T_{in} and T_{oc}, respectively.

2.3.2 *Single Channel Heat Coulomb Blockade*

The implications of Eq. (2.27) are most conveniently analyzed in two distinct limits: (i) a very cold output contact (OC) where the temperature $T_{\text{oc}} \to 0$, and (ii) a very cold input channel with $T_{\text{in}} \to 0$ and a fixed finite T_{oc}. Initially, note that for $T_p \to 0$, the equilibrium charge current noise power, as denoted by Eq. (2.22), simplifies to:

$$S_p(\omega) = \frac{\omega R_q^{-1}}{1 - e^{-\omega/T_p}} \to \omega R_q^{-1} \theta(\omega). \tag{2.30}$$

For scenario (i), by substituting the limit given in Eq. (2.30) for $p = \text{oc}$ into the definition of the effective temperature (2.26) and utilizing the asymptotic behaviors (2.29), we derive:

$$\frac{T_{\text{out}}}{T_{\text{in}}} = \begin{cases} \sqrt{1 - \frac{3\pi^2 \tau_C T_{\text{in}}}{5}} & \text{if } \tau_C T_{\text{in}} \ll 1, \\ \sqrt{\frac{3}{\pi \tau_C T_{\text{in}}}} & \text{if } \tau_C T_{\text{in}} \gg 1. \end{cases} \tag{2.31}$$

From Eq. (2.31), it is evident that for sufficiently small $\tau_C T_{\text{in}}$, the OC, despite its low temperature, does not effectively cool the input channel, essentially $T_{\text{in}} = T_{\text{out}}$. In contrast, for large $\tau_C T_{\text{in}}$, a relationship of $T_{\text{out}} \propto \sqrt{T_{\text{in}}/\tau_C}$ is established.

For configuration (ii), substituting the noise power (2.30) for $p = \text{in}$ into the effective output temperature (2.26), and applying the asymptotic limits (2.29), we find:

$$\frac{T_{\text{out}}}{T_{\text{oc}}} = \begin{cases} \sqrt{\frac{2}{5}}\pi \tau_C T_{\text{oc}} & \text{if } \tau_C T_{\text{oc}} \ll 1, \\ 1 - \frac{3}{2\pi \tau_C T_{\text{oc}}} & \text{if } \tau_C T_{\text{oc}} \gg 1. \end{cases} \tag{2.32}$$

This result indicates that the ability of the OC to heat the cold input is hindered for sufficiently small $\tau_C T_{\text{oc}}$, transitioning to $T_{\text{out}} \to 0$. In the regime where $\tau_C T_{\text{oc}}$ is large, we observe that $T_{\text{out}} \to T_{\text{oc}}$, which means efficient heating.

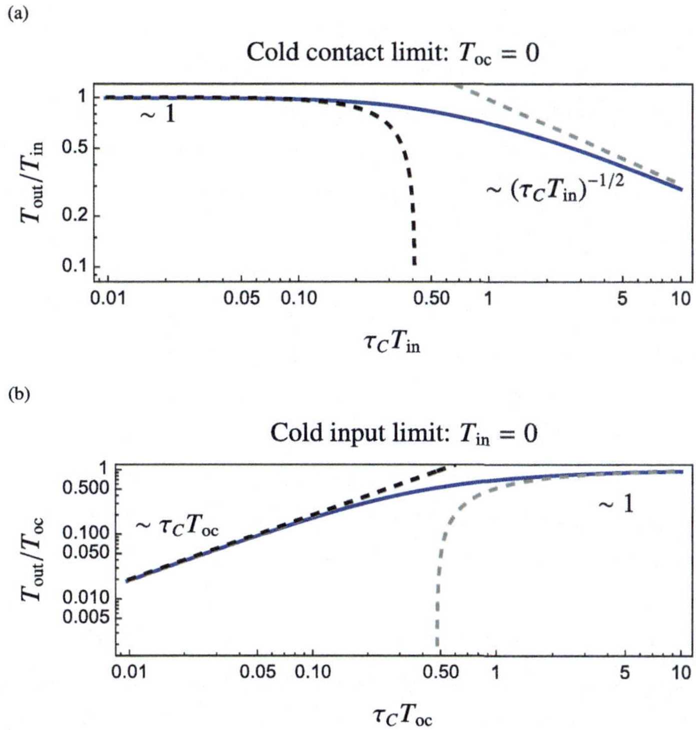

Fig. 2.2 **a** Blue solid line: Ratio of output and input effective temperatures, T_{out}/T_{in}, versus $\tau_C T_{in}$, where $\tau_C = R_q C$ is the RC time and T_{in} is the input channel temperature [22]. The dashed gray and black lines are the analytically obtained limits (2.31). The plots are shown in the limit of a cold OC, $T_{oc} = 0$. **b** Blue solid line: Ratio of output and OC effective temperatures, T_{out}/T_{oc}, versus $\tau_C T_{oc}$, where $\tau_C = RC$ is the RC time and T_{oc} is the OC temperature. The dashed gray and black lines are the analytically obtained limits (2.32). The plots are shown in the limit of a cold input channel, $T_{in} = 0$

The analytically derived limits in Eqs. (2.31) and (2.32), together with the exact expression of Eq. (2.27), are illustrated in Fig. 2.2a and b, respectively. These findings highlight the limited capacity of the OC to both cool and heat, which can be interpreted as an inability to fully equilibrate the incoming chiral edge channel. This feature is essentially a band-pass filtering effect in the quantum regime and lies at the heart of the heat Coulomb blockade effect [3], see also Sect. 2.4.1 below.

2.4 Generalization to Multiple Edge Channels

In the preceding section, our focus was restricted to configurations involving a single edge channel interfacing with the OC, as illustrated in Fig. 1.1. We now extend our analysis to scenarios featuring N incoming and N outgoing edge channels, depicted

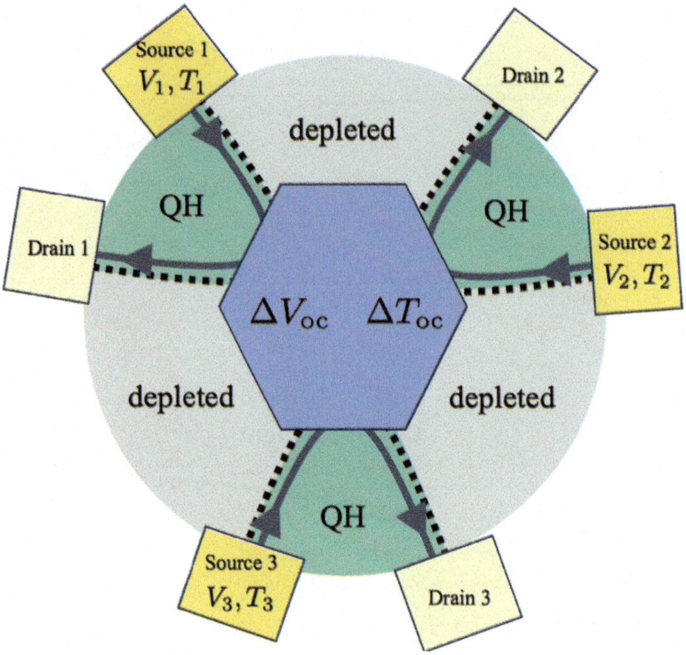

Fig. 2.3 Schematics of an Ohmic contact (OC, in blue) connected to multiple, chiral edge channels in the quantum Hall (QH) regime. Here, three pairs, $N = 3$, of edge channels are depicted. Pairs of incoming/outgoing edge channels appear on the boundaries (black, dotted lines) between QH bulk regions (turquoise regions, filling factor ν) and depleted regions (gray regions). Each pair is referred to as a "branch" in the main text. All input channels are fed from large metallic contacts (in yellow) characterized by input voltages, V_{in}, and input temperatures, T_{in}, for $m = 1, \ldots, N$. Fluctuations carried by the input edge channels generate voltage, ΔV_{oc}, and temperature, ΔT_{oc}, fluctuations. In turn, these quantities produce charge and heat fluctuations in the outgoing channels, which can be detected in the drain contacts (in light yellow)

in Fig. 2.3. Multichannel configurations have been instrumental in investigating the edge heat conductance in a broad range of quantum Hall states, as shown by studies such as [4, 13–20]. For the sake of simplicity, we assume that each connected edge channel is identical, characterized by uniform charge conductance. Nevertheless, our framework is adaptable to configurations involving heterogeneous channels, including those with non-Abelian edge modes.

2.4.1 *Multi-channel Charge Dynamics*

For the dynamics of charge in the scenario involving N incoming and outgoing edge channels, the Langevin (2.14b) and Kirchhoff equation (2.14a), generalize to the following in the N-channel case:

$$\frac{dQ(t)}{dt} = \sum_{m=1}^{N} \left[I_{\mathrm{in},m}(t) - I_{\mathrm{out},m}(t) \right], \tag{2.33a}$$

$$I_{\mathrm{out},m}(t) = \frac{Q(t)}{N\tau_{C,N}} + I_{\mathrm{oc},m}(t), \tag{2.33b}$$

where $Q(t)$ represents the charge that varies over time in the OC, and $I_{\mathrm{in},m}(t)$, $I_{\mathrm{out},m}(t)$, and $I_{\mathrm{oc},m}(t)$ denote the incoming, outgoing and thermal currents in branch m, respectively. A "branch" is defined as a pair of incoming and outgoing edge channels, as depicted in Fig. 2.3. The term $\tau_{C,N} \equiv \tau_C/N$ represents the *total* RC time of the OC for N attached channels. Following the approach used for the single-channel case in Sect. 2.3.2, we solve Eqs. (2.33a) and (2.33b) for the output currents, their fluctuations, and the associated noise powers, yielding:

$$I_{\mathrm{out},m}(\omega) = \sum_{p=\mathrm{in,oc}} \sum_{n=1}^{N} \left(\mathcal{T}_p(\omega)\right)_{mn} I_{p,n}(\omega), \tag{2.34a}$$

$$\delta I_{\mathrm{out},m}(\omega) = \sum_{p=\mathrm{in,oc}} \left(\mathcal{T}_p(\omega)\right)_{mn} \delta I_{p,n}(\omega), \tag{2.34b}$$

$$(\mathcal{T}_{\mathrm{oc}}(\omega))_{mn} = \delta_{mn} - \frac{1}{N}\left[1 - i\omega\tau_{C,N}\right]^{-1}, \tag{2.34c}$$

$$(\mathcal{T}_{\mathrm{in}}(\omega))_{mn} = \frac{1}{N}\left[1 - i\omega\tau_{C,N}\right]^{-1}, \tag{2.34d}$$

$$S_{\mathrm{out},m}(\omega) = \sum_{p=\mathrm{in,oc}} \sum_{n=1}^{N} \left|\left(\mathcal{T}_p(\omega)\right)_{mn}\right|^2 S_{p,n}(\omega). \tag{2.34e}$$

Assuming equal input charge noise powers for all channels, $S_{\mathrm{in},n}(\omega) = S_{\mathrm{in}}(\omega)$, and uniform OC contributions, $S_{\mathrm{oc},n}(\omega) = S_{\mathrm{oc}}(\omega)$, the output heat noise power (2.34e) simplifies to:

$$S_{\mathrm{out},m}(\omega) = \frac{S_{\mathrm{in}}(\omega) + S_{\mathrm{oc}}(\omega)\left[N - 1 + N\omega^2\tau_{C,N}^2\right]}{N(1 + \omega^2\tau_{C,N}^2)}, \tag{2.35}$$

indicating uniform charge noise power, channel number m independent of each output channel. At zero frequency $\omega = 0$, this reduces to:

$$S_{\mathrm{out},m}(0) = \frac{R_q^{-1}\left(T_{\mathrm{in}} + T_{\mathrm{oc}}(N-1)\right)}{N}, \tag{2.36}$$

revealing that, unlike the single-channel scenario in Fig. 1.1, the zero-frequency output noise in the multichannel setup gives access to the OC temperature T_{oc}, a parameter often used to deduce the heat conductance of edge channels [18].

Furthermore, we define the cross-correlation noise power for $m \neq n$ as:

$$\langle \delta I_{\text{out},m}(\omega) \delta I_{\text{out},n}(\omega') \rangle \equiv 2\pi \delta(\omega + \omega') S_{\text{out},m,n}(\omega), \tag{2.37}$$

and obtain

$$S_{\text{out},m,n}(\omega) = \frac{S_{\text{in}}(\omega) - S_{\text{oc}}(\omega)}{N(1 + \omega^2 \tau_{C,N}^2)}. \tag{2.38}$$

which, for zero frequency, simplifies to:

$$S_{\text{out},m,n}(0) = \frac{R_q^{-1}\,(T_{\text{in}} - T_{\text{oc}})}{N}. \tag{2.39}$$

By combining the expressions for the zero frequency auto- and cross-correlation output noise powers, we establish a sum rule reflecting charge conservation in the multichannel system. A comparison of the charge-noise powers at finite frequencies reveals the relationship:

$$S_{\text{out},m,n}(\omega) = S_{\text{out},m}(\omega) - S_{\text{oc}}(\omega). \tag{2.40}$$

Linking this with the N-channel Langevin equation (2.33b), the OC charge-correlation function, $\langle \Delta Q(\omega) \Delta Q(\omega') \rangle$, can be computed from the output noise powers as:

$$\langle \Delta Q(\omega) \Delta Q(\omega') \rangle = 2\pi \delta(\omega + \omega') \tau_{C,N}^2 \frac{S_{\text{in}}(\omega) + S_{\text{out},m}(\omega) - S_{\text{out},m,n}(\omega)}{N(1 + \omega^2 \tau_{C,N}^2)}. \tag{2.41}$$

In a single-channel setup ($N = 1$), separating the OC charge correlations from thermal current fluctuations is non-trivial.

2.4.2 Multi-channel Heat Coulomb Blockade

Starting from Eqs. (2.34a)–(2.34e), we demonstrate the heat Coulomb blockade effect in multichannel configurations, drawing on insights from Refs. [2, 3] and additional analysis Ch.5 in Ref. [21]. The essence of the heat Coulomb blockade is the suppression of a single quantum of measured heat conductance in setups with multiple channels. This phenomenon can be intuitively grasped by considering a transformation from the basis of N physically charged channels to a basis of one charged mode and $N - 1$ charge-neutral (dipole) modes. This transformation is not unique and transforms the original spatially local basis to a basis where charge and neutral modes are distributed non-locally across the branches in the device. However, it reveals that only the charged mode interacts with the OC capacitance, while the neutral modes do not. Hence, the neutral modes equilibrate efficiently with the OC and their outgoing heat currents are associated with the OC temperature, i.e. $T_{\text{out,neutral}} = T_{\text{oc}}$. In contrast, for $\tau_{C,N} T_{\text{oc}}, \tau_{C,N} T_{\text{in}} \ll 1$, the charged mode cannot equilibrate and $T_{\text{out,charged}} = T_{\text{in}}$, see Eq. (2.31). Hence, in the standard heat conductance setup, pioneered in Ref. [4], the charged mode contribution to the net heat current, see Eq. (2.42) below, disappears, since $J_{\text{charge}} \propto T^2_{\text{out,charged}} - T^2_{\text{in}} = 0$ for this mode, but for the neutral modes $J_{\text{neutral}} \propto T^2_{\text{oc}} - T^2_{\text{in}}$ which therefore do contribute to the net heat current. The total heat conductance, obtained by summing the contributions of all N channels (this sum is basis-independent due to charge conservation), is thus reduced by exactly one unit, which corresponds to the charged mode failing to equilibrate.

Let us now show this effect by explicit calculations. The key quantity of interest for heat Coulomb blockade is the net heat current, defined as

$$J_{\text{net}} = \frac{\kappa_0}{2} \sum_{n=1}^{N} (T^2_{\text{out},n} - T^2_{\text{in},n}) = \frac{\kappa_0}{2} \sum_{n=1}^{N} (T^2_{\text{out},n} - T^2_{\text{in}}), \tag{2.42}$$

assuming equal temperatures for all input channels, $T_{\text{in},n} = T_{\text{in}}$, in accordance with the typical experimental conditions [2, 3]. Next, we substitute the output temperatures $T^2_{\text{out,n}}$ calculated from noise powers (2.34e). Summing all these temperature contributions and plugging them into Eq. (2.42), we arrive at

$$J_{\text{net}} = N \times \frac{\kappa_0}{2} (T^2_{\text{oc}} - T^2_{\text{in}}) - \frac{1}{2\pi \tau^2_{C,N}} \left[\mathcal{F}\left(\frac{1}{\tau_{C,N} T_{\text{oc}}} \right) - \mathcal{F}\left(\frac{1}{\tau_{C,N} T_{\text{in}}} \right) \right], \tag{2.43}$$

with $\mathcal{F}(a)$ given in Eq. (2.28). Note that it is the N-channel RC-time $\tau_{C,N}$ that is included in Eq. (2.43).

Next, by using the asymptotic expressions (2.29), we see that for $\tau_{C,N} T_{\text{oc}} \ll 1$ and $\tau_{C,N} T_{\text{in}} \ll 1$, Eq. (2.43) reduces to

$$\begin{aligned} J_{\text{net}} &\approx N \times \frac{\kappa_0}{2} (T^2_{\text{oc}} - T^2_{\text{in}}) - \frac{1}{2\pi \tau^2_{C,N}} \left[\frac{\pi^2 \tau^2_{C,N} T^2_{\text{oc}}}{6} - \frac{\pi^2 \tau^2_{C,N} T^2_{\text{in}}}{6} \right] \\ &= (N-1) \times \frac{\kappa_0}{2} (T^2_{\text{oc}} - T^2_{\text{in}}). \end{aligned} \tag{2.44}$$

Hence, the net heat current is reduced by *exactly one unit*. We also see that for $N = 1$, the heat net current completely vanishes, consistent with the results in Sect. 2.3.2. In the opposite limit, $\tau_{C,N} T_{\text{oc}} \gg 1$ and $\tau_{C,N} T_{\text{in}} \gg 1$, Eq. (2.43) becomes instead

$$J_{\text{net}} \approx N \times \frac{\kappa_0}{2} (T^2_{\text{oc}} - T^2_{\text{in}}) - \frac{1}{4\tau_{C,N}} (T_{\text{oc}} - T_{\text{in}}), \tag{2.45}$$

i.e., a small correction per channel. The correction prefactor can alternatively be written in terms of the OC charging energy $E_C = \pi \tau_C^{-1}$ as

$$\frac{1}{4\tau_{C,N}} = \frac{NE_C}{4\pi}. \tag{2.46}$$

The correction term in Eq. (2.45) therefore increases linearly with increasing charging energy.

References

1. Duprez H, Pierre F, Sivre E, Aassime A, Parmentier FD, Cavanna A, Ouerghi A, Gennser U, Safi I, Mora C, Anthore A (2021) Dynamical Coulomb blockade under a temperature bias. Phys Rev Res 3:023122. https://doi.org/10.1103/PhysRevResearch.3.023122
2. Sivre E, Anthore A, Parmentier FD, Cavanna A, Gennser U, Ouerghi A, Jin Y, Pierre F (2018) Heat Coulomb blockade of one ballistic channel. Nat Phys 14:145–148. https://doi.org/10.1038/nphys4280
3. Sivre E, Duprez H, Anthore A, Aassime A, Parmentier FD, Cavanna A, Ouerghi A, Gennser U, Pierre F (2019) Electronic heat flow and thermal shot noise in quantum circuits. Nat Commun 10:5638. https://doi.org/10.1038/s41467-019-13566-8
4. Jezouin S, Parmentier FD, Anthore A, Gennser U, Cavanna A, Jin Y, Pierre F (2013) Quantum limit of heat flow across a single electronic channel. Science 342:601–604. https://doi.org/10.1126/science.1241912
5. le Sueur H, Altimiras C, Gennser U, Cavanna A, Mailly D, Pierre F (2010) Energy relaxation in the integer quantum Hall regime. Phys Rev Lett 105:056803. https://doi.org/10.1103/PhysRevLett.105.056803
6. Idrisov EG, Levkivskyi IP, Sukhorukov EV (2022) Thermal drag effect in quantum Hall circuits. Phys Rev B 106:L121405. https://doi.org/10.1103/PhysRevB.106.L121405
7. Matveev KA (1995) Coulomb blockade at almost perfect transmission. Phys Rev B 51:1743–1751. https://doi.org/10.1103/PhysRevB.51.1743
8. Furusaki A, Matveev KA (1995) Theory of strong inelastic cotunneling. Phys Rev B 52:16676–16695. https://doi.org/10.1103/PhysRevB.52.16676
9. Slobodeniuk AO, Levkivskyi IP, Sukhorukov EV (2013) Equilibration of quantum Hall edge states by an Ohmic contact. Phys Rev B 88:165307. https://doi.org/10.1103/PhysRevB.88.165307
10. Nyquist H (1928) Thermal agitation of electric charge in conductors. Phys Rev 32:110–113. https://doi.org/10.1103/PhysRev.32.110
11. Johnson JB (1928) Thermal agitation of electricity in conductors. Phys Rev 32:97–109. https://doi.org/10.1103/PhysRev.32.97
12. Idrisov EG, Levkivskyi IP, Sukhorukov EV (2025) Charge-conserving equilibration of quantum Hall edge states. Phys Rev Res 7:013042. https://doi.org/10.1103/PhysRevResearch.7.013042
13. Banerjee M, Heiblum M, Rosenblatt A, Oreg Y, Feldman DE, Stern A, Umansky V (2017) Observed quantization of anyonic heat flow. Nature 545:75–79. https://doi.org/10.1038/nature22052
14. Banerjee M, Heiblum M, Umansky V, Feldman DE, Oreg Y, Stern A (2018) Observation of half-integer thermal Hall conductance. Nature 559:205–210. https://doi.org/10.1038/s41586-018-0184-1
15. Srivastav SK, Sahu MR, Watanabe K, Taniguchi T, Banerjee S, Das A (2019) Universal quantized thermal conductance in graphene. Sci Adv 5:eaaw5798. https://doi.org/10.1126/sciadv.aaw5798
16. Srivastav SK, Kumar R, Spånslätt C, Watanabe K, Taniguchi T, Mirlin AD, Gefen Y, Das A (2021) Vanishing thermal equilibration for hole-conjugate fractional quantum Hall states in graphene. Phys Rev Lett 126:216803. https://doi.org/10.1103/PhysRevLett.126.216803

17. Srivastav SK, Kumar R, Spånslätt C, Watanabe K, Taniguchi T, Mirlin AD, Gefen Y, Das A (2022) Determination of topological edge quantum numbers of fractional quantum Hall phases by thermal conductance measurements. Nat Commun 13:1–8. https://doi.org/10.1038/s41467-022-32956-z
18. Melcer RA, Dutta B, Spånslätt C, Park J, Mirlin AD, Umansky V (2022) Absent thermal equilibration on fractional quantum Hall edges over macroscopic scale. Nat Commun 13:1–7. https://doi.org/10.1038/s41467-022-28009-0
19. Le Breton G, Delagrange R, Hong Y, Garg M, Watanabe K, Taniguchi T, Ribeiro-Palau R, Roulleau P, Roche P, Parmentier FD (2022) Heat equilibration of integer and fractional quantum Hall edge modes in graphene. Phys Rev Lett 129:116803. https://doi.org/10.1103/PhysRevLett.129.116803
20. Dutta B, Umansky V, Banerjee M, Heiblum M (2022) Isolated ballistic non-Abelian interface channel. Science 377:1198–1201. https://doi.org/10.1126/science.abm6571
21. Sivré É (2019) Electrical fluctuations and heat flow in a quantum composite circuit. PhD thesis, Université Paris Saclay. https://theses.hal.science/tel-02798535/
22. Spånslätt C, Stäbler F, Sukhorukov EV, Splettstoesser J (2024) Impact of potential and temperature fluctuations on charge and heat transport in quantum Hall edges in the heat Coulomb blockade regime. Phys Rev B 110(7):075431. https://doi.org/10.1103/PhysRevB.110.075431

Part II
Heat Transport and Dissipation in Quantum Systems

Chapter 3
The Transmission Line–Dissipation

Abstract This chapter presents the transmission line (TL) model as an extension of the Ohmic contact framework to describe dissipation in quantum Hall edge states. The TL model consists of a series of Ohmic contacts that act as reservoirs, partially equilibrating incoming currents and introducing dissipation. A key motivation for this model is the "missing heat paradox," where experimental measurements suggest a discrepancy between injected and detected heat flux. By incorporating Kirchhoff's laws and Langevin dynamics, we analyze charge and heat transport within the TL model, introducing a basis transformation that distinguishes between symmetric (charged) and dipole (neutral) modes. We demonstrate that despite dissipation, both modes carry a fully quantized heat flux at equilibrium, contrary to previous theoretical claims. This result follows from a generalized sum rule that ensures thermal equilibrium is maintained. Our findings challenge previous models predicting suppressed heat transport in neutral modes and suggest that dissipation alone cannot explain the missing heat paradox. We propose that deviations from quantized heat flux may stem from measurement effects rather than fundamental breakdowns in heat transport. The TL model provides a robust framework for exploring dissipation, nonlocal interactions, and electronic correlations in chiral quantum systems.

3.1 TL Approach for Chiral Systems with Intrinsic Dissipation

The Ohmic contact (OC) model, as described in Chap. 2, finds its application in the conceptualization of a transmission line (TL) model. This model utilizes a sequence of OCs, forming the transmission line, to effectively represent a quantum Hall edge state subject to disorder. The TL model is schematically represented in Fig. 3.1. Each OC, or node within this framework, is characterized by a self-capacitance C, with its dynamics governed by Kirchhoff's law (2.14a) and a Langevin equation (2.14b), or their extensions to multi-channel configurations Eqs. (2.33a) and (2.33b), respectively. Each node serves as a reservoir of infinitely many neutral degrees of freedom that are able to (partially) equilibrate incoming currents. Represented by Langevin sources, these neutral degrees of freedom are characterized by an equilibrium noise

F. Stäbler, *Dissipation, Interactions, and Heat Transport in Quantum Hall Devices*, Springer Theses, https://doi.org/10.1007/978-3-031-90671-8_3

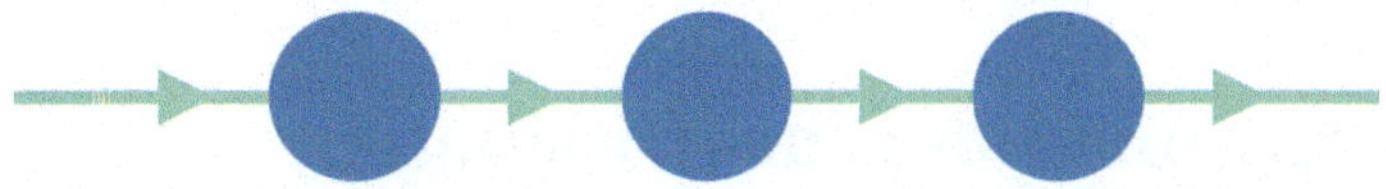

Fig. 3.1 A schematic of the transmission line (TL) model for quantum Hall edge states, where a chiral edge state, depicted in light blue, is interconnected with a series of Ohmic contacts (dark blue). Each OC introduces a self-capacitance C and is modeled by Kirchhoff's laws and Langevin dynamics to effectively model dissipation, disorder, and non-local interactions

power at a specified temperature, as described in Eq. (2.22). Using methods similar to those for the single OC, we can solve for the fluctuations in outgoing currents between these reservoirs. Subsequently, the heat flux between nodes can be calculated by integrating over the noise power, as shown in Eq. (2.26).

In the following, we present two variants of the transmission line model. The first variant, discussed here in Chap. 3, emphasizes the dissipation occurring within the nodes. The second variant, discussed in Chap. 4, focuses on the non-local interactions between nodes.

3.1.1 The "Missing Heat Paradox"

Our main motivation to explore the TL model, which incorporates intrinsic dissipation, is the *missing heat paradox*. Specifically, a discrepancy between the injected and measured heat flux (3.1), observed in a seminal experiment in quantum Hall (QH) systems at a filling factor $\nu = 2$ [4] and also [5, 6]; see Chap. 1 for a discussion of the experiment.

The authors of [4] concluded that the results of the experiment could be explained by the presence of additional degrees of freedom. However, for these observations to align with the experimental data, this additional mode would lead to a deviation from the established quantization principle of heat flux in one-dimensional systems. Specifically, to account for the observed discrepancy, this mode would have to transport an amount of heat less than a single heat flux quantum $J_q = \kappa_0/2\, T^2$. However, the experimentally measured heat flux

$$J_{\text{meas}} = \int \frac{d\varepsilon}{2\pi}\, \varepsilon \,(f(\varepsilon) - \theta(-\epsilon)) \tag{3.1}$$

deviates from the quantum flux of heat $J_{\text{meas}} \neq J_q$, especially when $f(\varepsilon)$ is not a Fermi-Dirac distribution, suggesting a potential violation of this paradigm.

In a previous work, [7] explored the theoretical possibility of a third neutral, co-propagating mode, known as the Aleiner-Glazman (AG) mode [8], serving as the necessary additional degree of freedom. This concept is associated with the electrostatic reconstruction of the QH edge [9, 10]. It was demonstrated that the AG excitation, due to dissipation, carries less than a quantum of heat flux. The

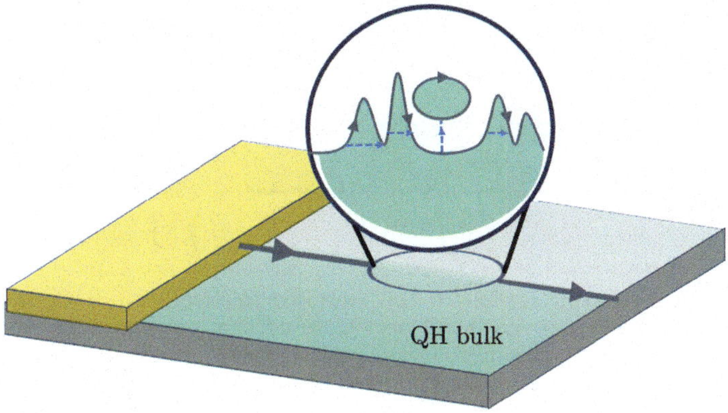

Fig. 3.2 The edge of QH bulk (in blue turquoise), showcasing a single edge state (marked by a gray arrow) emerging from a reservoir (in yellow) at a filling factor of $\nu = 1$. A detailed examination of the edge's microscopic structure is highlighted through the magnifying glass. It illustrates a potential dissipation mechanism within the QH edge, where imperfections may enable electron tunneling, effectively mimicking a basic electron reservoir. Such dynamics draw a parallel to a transmission line model, characterized by its distributed capacitors and resistors. Here, the electron reservoirs function similar to a heat bath, with temperatures determined in a self-consistent manner

charged mode in the system carries a full quantum of heat flux, $J_{\text{hydro}} = J_q$, with J_{hydro} defined as the energy flux in [7]. Subjecting the neutral mode to a transverse current, proportional to the longitudinal conductivity σ_{xx} within the compressible strip, introduces a dissipative component into its low-energy spectrum, effectively decreasing the heat it carries.

However, this analysis has notable limitations. First, the model applies only to low-energy degrees of freedom, characterized by wavelengths significantly exceeding the inverse of the compressible strip's size, ξ_{strip}^{-1}, due to the edge's inhomogeneity in the transverse direction. Despite this artificial cut-off, ξ_{strip}, the reduction in heat carried was found to be universal: $J_{\text{hydro}} = \frac{\sigma_{xx}}{2\pi\sigma_{xy}} J_q$. Second, the basis for these results, the heat flux operator J_{hydro}, was proposed as the symmetric product of the potential and current operators along the edge. The direct relationship between this operator, defined in the context of a two-dimensional hydrodynamic model J_{hydro}, and the measured heat flux (3.1)—which, for a linear spectrum, is given by the definition (2.16)—remains unclear. It should be noted that the observed heat loss might be attributed to non-local relaxation mechanisms [11–14].

The aim of this section is to develop a microscopically motivated effective model that emphasizes the concepts explored in [7], while eliminating the need for artificially introduced cut-offs and providing a clear definition of energy flux. The disordered nature of QH edges, characterized by lumps and puddles [15], which allow tunneling, can be conceptually understood as simple electron reservoirs, as schematically depicted in Fig. 3.2.

This comparison allows us to make an analogy with a transmission line comprised of distributed capacitors and resistors. In our model, disorder is represented through

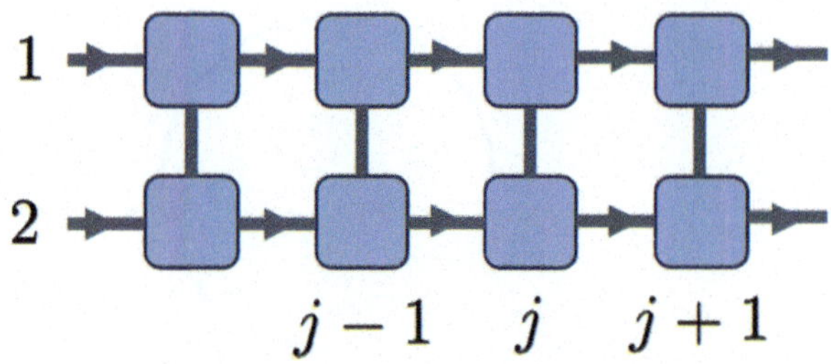

Fig. 3.3 The Transmission Line Model: This schematic represents two co-propagating modes consisting of multiple reservoirs capable of longitudinal and transverse interactions. Each reservoir within the system is distinctly identified by a positional index j and a label $\alpha = 1, 2$, where α denotes whether the reservoir belongs to the upper ($\alpha = 1$) or lower ($\alpha = 2$) channel of the transmission line

Ohmic contacts, and we employ a mix of Langevin equations and scattering states to incorporate dissipation within chiral systems [16]. By connecting a chiral system to a bath modeled as an open system, we gain the ability to explore both equilibrium and non-equilibrium scenarios analytically [1]. This methodological advantage positions our approach as a complementary technique to the Caldeira-Leggett model [17, 18], effectively capturing dissipation at the quantum level.

3.1.2 The TL Equations and Rotations

Let us start from a minimal model of filling factor $\nu = 2$ and model two co-propagating modes with a typically large (spin-) resistance between them (Fig. 3.3).[1]

We model the disorder at the edge of a Quantum Hall system as a simple metallic reservoir,[2] described by the Ohmic contacts introduced in the Chap. 2. The edge now consists of many OC nodes, interacting longitudinally (within the same channel) through a chiral quantum resistor of strength $R_q = 2\pi\hbar/e^2$, and transversely (between the channels) with a quantum resistor of, in principle, arbitrary strength $R_\perp$, as described in the equivalent circuit element shown in Fig. 3.4. The goal of this section is to compute the heat flux carried by the edge states in between the reservoirs

[1] Note that we consider in the present paper a model at filling factor $\nu = 2$ and explicitly model a (spin-flip) resistance between the two modes. However, this is completely analogue to modeling the hydrodynamic edge reconstruction, where the additional mode is only half filled. Our results can be simply connected to the hydrodynamic model in [7] by the transformation $R_q \rightarrow R_q/2$.

[2] Modeling the edge defects as an OC implies that the disorder has a level spacing much smaller than the charging energy. In the transmission line chapters, we set $\tau_E \rightarrow \infty$. For a discussion of relevant energy scales, see Sect. 2.1. This assumption must be experimentally verified; however, the opposite limit, where the level spacing of the OC is not negligible compared to the charging energy, remains an open question and poses intriguing theoretical challenges to be addressed in future work. One way to proceed could be to consider two-level systems and treat the tunneling to disorder perturbatively.

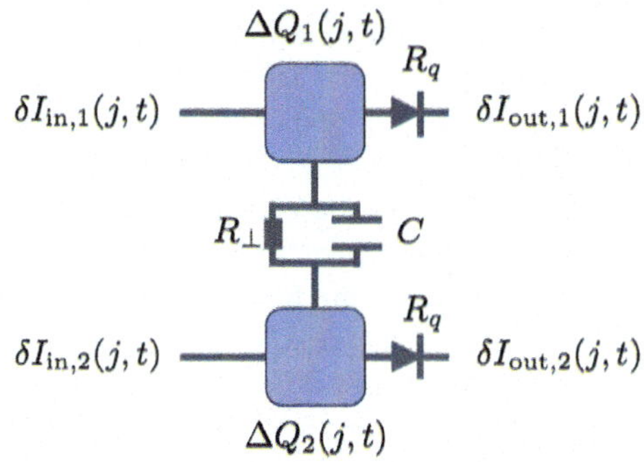

Fig. 3.4 A transverse cross-section of the transmission line depicted in Fig. 3.3. The Ohmic reservoirs longitudinally emit currents according to Ohm's law with resistance R_q plus thermal fluctuations. The upper and lower modes interact via a transverse resistor $R_\perp$. The capacitor represents the self-capacitance C of the Ohmic contacts

and to check how the heat flux is affected by the dissipation within the OCs and by $R_\perp$.

Building on Chap. 2, we adopt the Langevin formalism (2.14b) and Kirchhoff's law (2.14a) for the TL segment shown in Fig. 3.4, introducing two chains of N nodes interacting via a transverse resistor. This leads to the *transmission line equations*:

$$\frac{d}{dt}\Delta Q_\alpha(j) = \delta I_{\text{in},\alpha}(j) - \delta I_{\text{out},\alpha}(j) - \text{sgn}(\alpha)\delta I_{\text{diss}}(j), \tag{3.2a}$$

$$\delta I_{\text{out},\alpha}(j) = \frac{1}{\tau_C}\Delta Q_\alpha(j) + \delta I_{\text{oc},\alpha}(j) \tag{3.2b}$$

$$\delta I_{\text{diss}}(j) = \frac{\Gamma}{2\tau_C}(\Delta Q_1(j) - \Delta Q_2(j)) + \delta I_\perp(j), \tag{3.2c}$$

$$\delta I_{\text{in},\alpha}(j) = \delta I_{\text{out},\alpha}(j-1), \tag{3.2d}$$

where j denotes the position of the node, $\alpha = 1, 2$ indicates the upper or lower channel, $\text{sgn}(\alpha) = +1$ for $\alpha = 1$ and -1 for $\alpha = 2$, and $\Gamma = 2R_q/R_\perp C$ represents the relative strength of the perpendicular resistor. The Langevin sources at each node, $\delta I_{\text{oc},\alpha}(j, t)$ and $\delta I_\perp(j, t)$, are assumed to be in thermal equilibrium, characterized by an equilibrium noise power given by Eq. 2.22:

$$\langle \delta I_p(\omega)\delta I_{p'}(\omega') \rangle = 2\pi\delta_{pp'}\delta(\omega + \omega')S_p(\omega) \tag{3.3a}$$

$$S_p(\omega) = \frac{\omega R_q^{-1}}{1 - e^{-\omega/T_p}}, \quad p \in \{\text{oc}, \perp\}. \tag{3.3b}$$

This chapter assumes identical temperatures across the transmission line, so the noise power has no spatial dependence and channel number dependence, $S_{\text{oc},\alpha}(j, \omega) = S_{\text{oc}}(\omega)$. We only explore the equilibrium case where $T_{\text{oc}} = T_\perp$, noting that our formalism is also able to tackle nonequilibrium scenarios and investigate laws such as the Wiedemann-Franz law or Fourier's law of heat conduction. These analyses extend beyond the scope of this chapter and are subject for future investigation; see Chap. 7 for ideas.

Next, we change the basis of two charged, physical channels into one with a symmetric channel "s" (with conductance $2/R_q$) and one dipole channel "a" (or "dipole mode"). This transformation is not unique; whereas channels in the original basis are spatially local, symmetric, and dipole modes distribute non-locally across the device's branches. The *rotated transmission line equations* are as follows:

$$\frac{d}{dt}\Delta Q^s(j) = \delta I^s_{\text{out}}(j-1) - \delta I^s_{\text{out}}(j), \tag{3.4a}$$

$$\frac{d}{dt}\Delta Q^a(j) = \delta I^a_{\text{out}}(j-1) - \delta I^a_{\text{out}}(j) - \delta I_{\text{diss}}(j), \tag{3.4b}$$

$$\delta I^{s/a}_{\text{out}}(j) = \frac{1}{\tau_C}\Delta Q^{s/a}(j) + \delta I^{s/a}_{\text{oc}}(j) \tag{3.4c}$$

$$\delta I_{\text{diss}}(j) = \frac{\Gamma}{\tau_C}\Delta Q^a(j) + \delta I_\perp(j), \tag{3.4d}$$

highlighting that only the dipole mode experiences the dissipation introduced by the transverse resistor. We solve the *rotated transmission line equations* (3.4a)–(3.4d) using a discrete Fourier transform given by

$$X(j,t) = \sum_{k=0}^{N-1}\int \frac{d\omega}{2\pi} e^{i\frac{2\pi k}{N}j - i\omega t} X(k,\omega), \tag{3.5a}$$

$$X(k,\omega) = \frac{1}{N}\sum_{j=0}^{N-1}\int dt e^{-i\frac{2\pi k}{N}j + i\omega t} X(j,t), \tag{3.5b}$$

It is important to note that the application of the Fourier transform assumes periodic boundary conditions, yet the presence of dissipative terms in the spectra of symmetric and dipole modes implies that boundary conditions become irrelevant in the thermodynamic limit $N \to \infty$ we apply later on. The spectrum of the symmetric mode is

$$-i\omega^s_k = (\tau_C)^{-1}\left(e^{-2\pi i k/N} - 1\right), \tag{3.6}$$

and of the dipole mode

$$-i\omega^a_k = (\tau_C)^{-1}\left(e^{-2\pi i k/N} - 1 - \Gamma\right), \tag{3.7}$$

which both contain highly nonlinear terms responsible for dissipation and dispersion of the modes. The dipole mode contains furthermore an additional imaginary part governed by the strength of the perpendicular resistor $\Gamma = 2R_q/R_\perp$. Note that after linearizing the spectrum our approach becomes equivalent to the hydrodynamic equations considered in [7].

3.1.3 Heat Flux of the Charged and Neutral Mode

One advantage of the TL approach is that the heat flux can be unambiguously defined as the heat flux carried by the edge states in between the nodes. The connecting edge states of the TL are free chiral edge states for which we can apply the usual definition of heat flux (2.16) and (2.26). The average heat flux between the nodes, carried by symmetric and dipole modes, is expressed as:

$$J = \frac{R_q}{2} \sum_{\alpha=1,2} \left\langle \left(\delta I_{\mathrm{out},\alpha}(j,t)\right)^2 \right\rangle = R_q \sum_{\beta=s,a} \left\langle \left(\delta I^{\beta}_{\mathrm{out}}(j,t)\right)^2 \right\rangle. \tag{3.8}$$

Solving the *rotated transmission line equations* (3.4a)–(3.4d) using a discrete Fourier transform allows to find the noise power that enters the heat flux[3] per crossection (3.8) in terms of the Langevin sources with a known noise power (3.3a).

We evaluated the contributions of the symmetric (J^s) and dipole (J^a) modes to the total heat flux separately. Importantly, the contribution of the symmetric mode can be directly inferred by taking the limit of $\Gamma \to 0$ closer to zero in J^a. We take the following steps: (i) Extending the number of nodes (N) to infinity, which replaces the k summation by the k integration while maintaining a fixed distance (ξ) between the nodes, playing the role of the disorder length scale. This can be viewed as a phenomenological input to this model; (ii) perform the k integral by projecting it onto the unit circle ($z \to e^{ik\xi}$) and applying the residue theorem for the enclosed poles; (iii) execute the ω-integral; see Appendix A for further details. The calculations yield the following results:

For the symmetric mode:

$$J^s = \frac{R_q}{2} \int \frac{d\omega}{2\pi} \int_{-\pi/\xi}^{\pi/\xi} \frac{dk}{2\pi} \frac{\xi \omega^2 S^s_{\mathrm{oc}}(\omega)}{(\omega - \omega^s(k))(\omega + \omega^s(-k))} = J_q, \tag{3.9}$$

where $S^s_{\mathrm{oc}}(\omega) = S(\omega)/2$ with the standard noise power (2.22). This demonstrates that the heat flux quantization ($J^s = J_q$) is fulfilled for the symmetric mode. Surprisingly,[4] for the dipole mode:

$$J^a = \frac{R_q}{2} \int \frac{d\omega}{2\pi} \int_{-\pi/\xi}^{\pi/\xi} \frac{dk}{2\pi} \frac{\xi \left(\left(\omega^2 \tau_C^2 + \Gamma^2\right) S^a_{\mathrm{oc}}(\omega) + S_{\perp}(\omega)\right)}{\tau_C^2 (\omega - \omega^a(k))(\omega + \omega^a(-k))} = J_q, \tag{3.10}$$

[3] Note that we assume in the definition of Eq. (3.8) that vacuum fluctuations are subtracted according to Eq. (2.26).

[4] In contrast to [7].

with the remarkable implication that, the neutral mode also contributes a full quantum of heat flux, unaffected by $R_\perp$. This holds at equal temperatures ($T_\perp = T_{\mathrm{oc}}$) using $S_\perp(\omega) = 2\Gamma S_{\mathrm{oc}}(\omega)$ and $S^a_{\mathrm{oc}}(\omega) = S(\omega)/2$.

This perfect quantization emerges after computing the momentum integral in Eq. (3.10), which makes the expression equivalent to Eq. (2.25), a step explained in detail in Appendix A. Such an outcome challenges initial expectations, as the neutral mode's dispersion relation is susceptible to arbitrarily strong dissipation, e.g. the strongly damped limit $\Gamma \to \infty$. In summary, our analysis reveals that collective mode dissipation does not violate the paradigm of equilibrium quantization of heat flux. This is one of the central results of this section.

3.1.4 The Sum Rule

To understand the reason for this perfect quantization, we closely inspect the solutions of the transmission line Eqs. (3.2)–(3.2c). In single channel OC, as discussed in Chap. 2, the condition of thermal equilibrium is guaranteed by the sum rule (2.23), first described by [19].

In the current scenario, we solve the transmission line equations for the outgoing current, $\delta I_{\mathrm{out},\alpha}(j,t)$, as a function of the incoming current, $\delta I_{\mathrm{in},\alpha}(j,t)$, and the Langevin sources $\delta I_{\mathrm{oc},\alpha}(j,t)$ and $\delta I_\perp(j,t)$. The current fluctuations in channels $\alpha \in \{1,2\}$ are given by

$$\delta I_{\mathrm{out},\alpha}(j,\omega) = \sum_{\substack{\beta=1,2,\\ p=\mathrm{in},\mathrm{oc}}} T_p^{\alpha\beta}(\omega)\delta I_{p,\beta}(j,\omega) + T_\perp^\alpha(\omega)\delta I_\perp(j,\omega). \tag{3.11}$$

The corresponding coefficients are given by

$$T_{\mathrm{in}}^{\alpha\alpha} = \frac{1}{2}\left(A(\omega) + B(\omega)\right), \tag{3.12a}$$

$$T_{\mathrm{in}}^{\alpha\beta} = \frac{1}{2}\left(A(\omega) - B(\omega)\right), \tag{3.12b}$$

$$T_\perp^2 = B(\omega), \tag{3.12c}$$

with $T_\perp^2 = -T_\perp^1$, $T_{\mathrm{in}}^{\alpha\beta} = -T_{\mathrm{oc}}^{\alpha\beta}$, $T_{\mathrm{oc}}^{\alpha\alpha} = 1 - T_{\mathrm{in}}^{\alpha\alpha}$, $A(\omega) = (1 - i\tau_C\omega)^{-1}$ and $B(\omega) = (1 + \Gamma - i\tau_C\omega)^{-1}$. When calculating the noise power in the upper/lower channel α, one finds that

$$S_{\mathrm{out},\alpha}(j,\omega) = \sum_{\substack{\beta=1,2,\\ p=\mathrm{in},\mathrm{oc}}} \left|T_p^{\alpha\beta}(\omega)\right|^2 S_p(j,\omega) + \left|T_\perp^\alpha(\omega)\right|^2 S_\perp(j,\omega). \tag{3.13}$$

The *generalized sum rule* states that if the current incident to a node is in equilibrium, that is, it has a noise power $S_{\text{in}}(j,\omega)$ as given by Eq. (3.3a), with temperatures $T_{\text{in}} = T_{\text{oc}} = T_{\perp}$ equal, we find that

$$\sum_{\substack{\beta=1,2,\\ p=\text{in},\text{oc}}} \left|T_p^{\alpha\beta}(\omega)\right|^2 G_p = \frac{1}{R_q}, \tag{3.14}$$

with $G_{\text{in}} = G_{\text{oc}} = \frac{1}{R_q}$ and $G_{\perp} = \frac{2}{R_{\perp}}$, thus generalizing the sum rule to the transmission line and guaranteeing an equilibrium output current despite the presence of dissipation. This is because the transmission line is energetically floating. However, in Paper II, we demonstrate how to break the sum rule in multichannel models with nonlocal couplings or in the presence of diffusive currents.

3.1.5 Exact Cancellation of Joule Heating and Backaction in Equilibrium

The structure of the outgoing current is always of the form

$$\delta I_{\text{out}} = \frac{1}{\tau_C}\Delta Q(t) + \delta I_{\text{oc}}(t), \tag{3.15}$$

hence the total current-current correlation function in Eq. (3.8) consists of three parts. The first contribution contains the autocorrelation function of the collective mode C_{QQ}, the second the cross-correlation function between the contribution of the collective mode and the Langevin sources C_{IQ}, and lastly, the autocorrelation function of the Langevin sources C_{II}. We discover the remarkable result:

$$\frac{1}{\tau_C}\int \frac{dk}{2\pi} C_{QQ} = -\int \frac{dk}{2\pi}\left(C_{IQ} + C_{QI}\right), \tag{3.16a}$$

$$C_{XY} = \langle \delta X^{\alpha}(k,\omega)\delta Y^{\alpha}(-k,-\omega)\rangle, \tag{3.16b}$$

which applies to both the symmetric and dipole modes, $\alpha = s, a$, and $X, Y \in \{\Delta Q, \delta I\}$. The exact cancelation after performing the momentum integral implies that the Joule heating on the nodes, given by C_{QQ}, is fully offset by the back action of the sources on the nodes, $\left(C_{IQ} + C_{QI}\right)$. The correlation function of the source, C_{II}, is the only remaining part of Eq. (3.8), which explains why there is always a quantum flux for the symmetric and dipole modes. Therefore, sources are not merely auxiliary but integral physical entities of the system. This principle of exact cancelation due to the back action of thermal noise, first introduced in Nyquist's work on thermal fluctuations in 1928 [2, 20], indicates that two resistors in thermal equilibrium connected by ideal wires generate thermal fluctuations. These fluctuations theoretically lead to a heat flux from one resistor to another but are balanced by the fluctuations of the second resistor. This balance is maintained in every frequency window since otherwise it would be possible to extract energy by using a frequency filter in the system. Thus, the cancelation of the correlation functions as shown in Eq. (4.2a) holds before the integration over frequencies. The simple reason behind the exact cancelation is the second law of thermodynamics, which prohibits heat extraction, or in this context, dissipation, if the system is in thermal equilibrium and isolated. This contrasts with the approach of Goremykina et al., 2019, where the energy flux was defined as the potential energy flux proportional to $\mathcal{C}_{QQ}$, a factor now nullified by the back-action effect of the sources. The implications of this will be further discussed in the following section.

3.1.6 Low-Energy Theory of an Edge with Intrinsic Dissipation

In this section, we highlight the differences between the hydrodynamic model presented in Goremykina et al., 2019, and the transmission line approach outlined in this chapter, explaining why the former leads to different outcomes despite the correct application of fluctuation-dissipation relations, a conventional method for deriving equilibrium correlation functions.

The core objective now is to derive a low-energy theory from the discrete transmission line model, compare it with the findings of the preceding study, and evaluate its universality. In the low-energy limit, particularly at low temperatures, not all modes within the non-linear spectrum of symmetric and dipole modes are excited. This allows for expansion of the spectrum, if possible, to the point where the heat flux integrals Eqs. (3.9) and (3.10) converge and give a quantum heat flux. Essentially, this process is equivalent to establishing low-energy field theory that describes chiral heat transport at the QH edge in the presence of dissipation. We will examine the symmetric and dipole modes in detail separately.

3.1.6.1 Low-Energy Field Theory for the Symmetric Mode

The dispersion relation of the symmetric mode is described by Eq. (3.6), revealing a "hidden" type of dispersion that emerges from the discreteness of the transmission line. This can be seen, when the dispersion relation is expanded for small values of ξ:

$$-i\omega^s(k) \approx -ikv_q - \frac{k^2\xi v_q}{2} + i\frac{k^3\xi^2 v_q}{6}, \tag{3.17}$$

where we have scaled to intensive quantities by introducing the velocity $v_q = \xi/\tau_C$. The second term describes dissipation, which is derived from the retardation of the collective mode, and disappears in the true continuum limit ($\xi \to 0$). It is important to note that this continuum limit is non-trivial; physically, this signifies that the decay length of the collective mode substantially exceeds the distance between nodes, resulting in no dissipation and thus no heating or back action effect, as discussed in the preceding section. In this scenario, the system's response becomes highly sensitive to its boundary conditions, and the potential energy flux, as outlined in [7], accurately predicts a quantum flux. Maintaining dissipation within the system and defining the heat flux through the current-current correlation function yields a heat flux quantum for low energies; the supplemental material offers detailed calculations. This allows us to represent the equation of motion in a coarse-grained manner, that is, dependent on a continuous variable rather than a discrete node index. This can be seen as the minimal expansion[5] of the discrete difference operator in Kirchhoff's law to correctly capture the characteristics of dissipation at low energies, complemented by a Langevin equation. In real space, the equations of motion for the coarse-grained collective mode $Q^s(x,t)$ and source $\delta I^s(x,t)$ are expressed as follows:

$$\delta I^s_{\text{out}}(x,t) = \frac{1}{\tau_C} \Delta Q^s(x,t) + \delta I^s_{\text{oc}}(x,t), \tag{3.18a}$$

$$\frac{1}{\tau_C} \partial_t \Delta Q^s(x,t) + v_q \mathcal{D} \Delta I^s_{\text{out}}(x,t) = 0 \tag{3.18b}$$

where $\mathcal{D} = \partial_x - \frac{\xi}{2}\partial_x^2 + \frac{\xi^2}{6}\partial_x^3$. Employing these modified hydrodynamic equations, we should: (i) Solve for the collective mode contribution ΔQ^s; (ii) Compute the heat flux carried by the chiral system as $J = R_q \left\langle \left(\Delta I^s_{\text{out}}(x,t) \right)^2 \right\rangle$; (iii) Take into account low energies and momenta by expanding the cubic and quartic terms in the momenta around $k \to \omega / v_q$, which will result in a quantum heat flux, as detailed in Appendix A.

3.1.6.2 Low-Energy Field Theory for the Dipole Mode

The dispersion relation of the dipole mode Eq. (3.7) contains two sources of dissipation: one due to the discreteness of the transmission line, similar to the symmetric mode, and the other resulting from the transverse resistor. Concentrating solely on the transverse dissipation, we consider only the leading-order terms:

$$-i\omega^a(k) \approx -i v_q k - \frac{v_q \Gamma}{\xi}. \tag{3.19}$$

As explicitly demonstrated in Appendix A, if $\Gamma \approx k\xi \ll 1$, the heat integral results in a quantum flux. If this condition is not fulfilled, the main contribution of the integral does not come from low momenta but from the system's bandwidth, necessitating integration over all of the spectrum's non-linearities. This indicates that the dipole

---

[5] Taking less terms does not give the correct heat flux.

collective mode decays over distances shorter than the node spacing under strong dissipation and is effectively pinned, leaving no low-energy theory for the strongly overdamped dipole mode. Nevertheless, thermal fluctuations continue to carry quantum flux. For cases where a low-energy theory is applicable, we can formulate a coarse-grained equation of motion[6]:

$$\delta I^a_{\text{out}}(x,t) = \frac{1}{\tau_C}\delta Q^a(x,t) + \delta I^a_{\text{oc}}(x,t), \tag{3.20a}$$

$$\delta I_{\text{diss}}(x,t) = \frac{\Gamma}{\tau_C}\Delta Q^a(x,t) + \delta I_\perp(x,t), \tag{3.20b}$$

$$\frac{1}{\tau_C}\partial_t \Delta Q^a(x,t) + v_q \partial_x \delta I^a_{\text{out}}(x,t) - \frac{1}{\tau_C}\delta I_{\text{diss}}(x,t) = 0. \tag{3.20c}$$

This framework allows for a discussion on the role of the artificial cutoff ξ_{strip} introduced in [7], for momentum integration, given by the transverse size of the compressible strip, a limitation of the low-energy approach to the edge. If the main contribution to the heat integral in [7] comes from small energies and momenta, replacing the artificial cutoff with the actual one may alter the numerical prefactor but does not affect the parametric suppression observed in the heat flux of the neutral mode. It is important to stress that the conceptual error lies in defining the heat flux as the potential energy flux. The Langevin sources presented in our formalism are not merely mathematical constructs but integral components of the system, leading to an autocorrelation cancelation of the collective mode. The Joule heating is given back by the sources' back-action on the collective mode in thermal equilibrium. Consequently, a Langevin equation represents a more comprehensive approach to dissipative systems, offering a more accurate definition of the energy flux that is unattainable from a hydrodynamic perspective.

3.1.7 *The Role of Aleiner-Glazman Modes*

Thus far, we have successfully written down an effective model for a system characterized by two co-propagating modes. This configuration describes a filling factor $\nu = 2$, accompanied by a transverse spin resistance. Alternatively, after rescaling ($R_q \to R_q/2$), this model parallels the hydrodynamic framework that contains one charged and one half-filled Aleiner-Glazman neutral mode. The latter comes from an approximation of the density profile produced by electrostatic analysis of the edge, as detailed in [8].

In developing an effective model for dissipation within these systems, we find that formulating a low-energy theory for the neutral mode is fundamentally flawed in the presence of dissipation. The inability to accurately derive the continuum limit

---

[6] The heat flux carried by the dipole mode may be evaluated in the same way as for the symmetric mode.

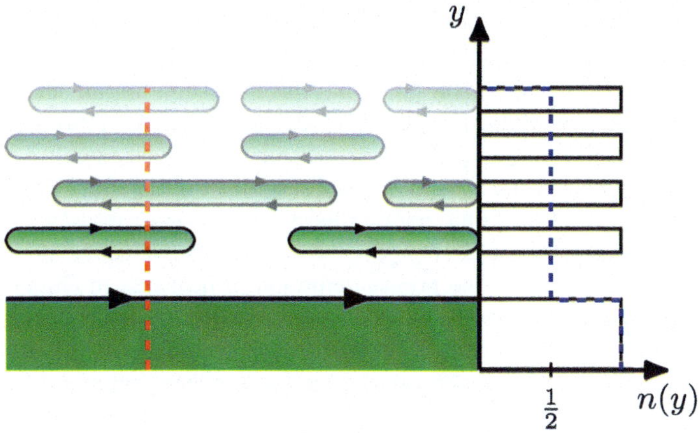

Fig. 3.5 This figure illustrates an effective depiction of edge defects represented by quantum Hall puddles, with the corresponding density profile shown in black, cutting through the red dashed line. The blue dashed line indicates the "average" hydrodynamic density, hinting at the existence of a half-filled neutral mode. However, the model introduced in this paper incorporates the disorder length scale into the hydrodynamic equations of motion, challenging the interpretation of the non-homogeneous density profile along the edge, attributed to quantum puddles, as simply an on average half-filled neutral mode. It suggests that if such a neutral mode does exist, it would decay over length scales similar to the disorder length scale, thereby questioning the conventional understanding of edge state homogeneity in the presence of quantum puddles

of the TL model, considering the edge reconstruction of the Aleiner-Glazman modes, highlights this issue. Additionally, the dissipation observed in the symmetric mode is dependent on length scales, such as the correlation length of disorder, which the low-energy approach fails to encapsulate. Consequently, we conclude that the electrostatic/hydrodynamic framework fails to capture dissipative chiral transport in Quantum Hall (QH) systems, while also overlooking empirical evidence observed at the QH edge. The conceptualization of QH puddles, as referred to in Fig. 3.5, is inconsistent with the theoretical predictions associated with ALG modes. The perceived half-filling emerges as an artifact of averaging numerous puddles, each exhibiting both co- and counterpropagating modes, which significantly differs from a single, half-filled co-propagating mode.

3.1.8 Concluding Remarks

We have demonstrated that the quantization of heat flux in chiral systems at thermal equilibrium is a robust phenomenon that underpins fundamental thermodynamic principles. Our analysis of chiral quantum systems used the "transmission line approach," which is based on the Langevin formalism and the scattering state formalism introduced in previous studies [16, 19]. This approach incorporates dissipation at an effective level. Drawing parallels with the Caldeira-Leggett model, where a bath of numerous oscillators is connected to a chiral system, we treat these baths as open systems. This enables the analysis of both equilibrium and nonequilibrium scenarios. Beyond addressing the observed reduction of heat flux in the strongly overdamped neutral mode claimed in prior research [7], we have established the quantization of heat flux in chiral systems in the presence of dissipation. We have linked the heat flux quantization to the local unitarity condition of the scattering matrix and the back-action effect of the Langevin sources on the collective mode. Through this work, we have clarified the proper definition of heat flux in chiral dissipative systems and discussed the transition to the potential energy flux concept utilized in previous studies [7]. This unveils the Langevin sources as genuine physical components capable of reintroducing heat into the system, rather than mere mathematical constructs. Further, we have advanced low-energy theoretical models derived from the transmission line framework that capture the universal features of the edge. Our discussion extends to the minimal expansion of Kirchhoff's law, rounding off the low-energy theory formulation. Although the results of this project do not resolve the "missing heat paradox", they open a field containing many interesting ideas; cf. [1]. A potential explanation for the experimental observations may lie in a discrepancy between actual and measured heat flux, a hypothesis that will be explored in subsequent chapters.

To proceed further, one needs to construct an electronic operator that goes beyond the traditional bosonization formalism by accounting for correlations through the collective mode when a tunnel probe is attached. This was done in paper III. Additionally, our findings indicate that the strongly overdamped AG modes fail the low-energy theoretical description and, if present, cannot account for the experimental data, leaving their experimental detection as an unresolved challenge. The formalism developed in this project can be easily applied to different edge reconstructions and to fractional quantum Hall systems.

References

1. Idrisov EG, Levkivskyi IP, Sukhorukov EV (2025) Charge-conserving equilibration of quantum Hall edge states. Phys Rev Res 7:013042. https://doi.org/10.1103/PhysRevResearch.7.013042
2. Nyquist H (1928) Thermal agitation of electric charge in conductors. Phys Rev 32:110–113. https://doi.org/10.1103/PhysRev.32.110
3. Johnson JB (1928) Thermal agitation of electricity in conductors. Phys Rev 32:97–109. https://doi.org/10.1103/PhysRev.32.97
4. le Sueur H, Altimiras C, Gennser U, Cavanna A, Mailly D, Pierre F (2010) Energy relaxation in the integer quantum Hall regime. Phys Rev Lett 105:056803. https://doi.org/10.1103/PhysRevLett.105.056803
5. Granger G, Eisenstein JP, Reno JL (2009) Observation of chiral heat transport in the quantum Hall regime. Phys Rev Lett 102:086803. https://doi.org/10.1103/PhysRevLett.102.086803
6. Venkatachalam V, Hart S, Pfeiffer L, West K, Yacoby A (2012) Local thermometry of neutral modes on the quantum Hall edge. Nat Phys 8:676–681. https://doi.org/10.1038/nphys2384
7. Goremykina A, Borin A, Sukhorukov E (2019) Heat current in a dissipative quantum Hall edge. arXiv:1908.01213

8. Aleiner IL, Glazman LI (1994) Novel edge excitations of two-dimensional electron liquid in a magnetic field. Phys Rev Lett 72:2935–2938. https://doi.org/10.1103/PhysRevLett.72.2935
9. Chklovskii DB, Shklovskii BI, Glazman LI (1992) Electrostatics of edge channels. Phys Rev B 46:4026–4034. https://doi.org/10.1103/PhysRevB.46.4026
10. de Chamon CC, Wen XG (1994) Sharp and smooth boundaries of quantum Hall liquids. Phys Rev B 49:8227–8241. https://doi.org/10.1103/PhysRevB.49.8227
11. Rodriguez RH, Parmentier FD, Ferraro D, Roulleau P, Gennser U, Cavanna A, Sassetti M, Portier F, Mailly D, Roche P (2020) Relaxation and revival of quasiparticles injected in an interacting quantum Hall liquid. Nat Commun 11:2426. https://doi.org/10.1038/s41467-020-16331-4
12. Lunde AM, Nigg SE (2016) Statistical theory of relaxation of high-energy electrons in quantum Hall edge states. Phys Rev B 94:045409. https://doi.org/10.1103/PhysRevB.94.045409
13. Fischer SG, Park J, Meir Y, Gefen Y (2019) Interaction-induced charge transfer in a mesoscopic electron spectrometer. Phys Rev B 100:195411. https://doi.org/10.1103/PhysRevB.100.195411
14. Krähenmann T, Fischer SG, Röösli M, Ihn T, Reichl C, Wegscheider W, Ensslin K, Gefen Y, Meir Y (2019) Auger-spectroscopy in quantum Hall edge channels and the missing energy problem. Nat Commun 10:3915. https://doi.org/10.1038/s41467-019-11888-1
15. Shamim S, Shekhar P, Beugeling W, Böttcher J, Budewitz A, Mayer J-B, Lunczer L, Hankiewicz EM, Buhmann H, Molenkamp LW (2022) Counterpropagating topological and quantum Hall edge channels. Nat Commun 13:2682. https://doi.org/10.1038/s41467-022-29815-2
16. Sukhorukov EV (2016) Scattering theory approach to bosonization of non-equilibrium mesoscopic systems. Physica E 77:191–198. https://doi.org/10.1016/j.physe.2015.11.018
17. Feynman RP, Vernon FL (1963) The theory of a general quantum system interacting with a linear dissipative system. Ann Phys 24:118–173. https://doi.org/10.1016/0003-4916(63)90068-X
18. Caldeira AO, Leggett AJ (1983) Path integral approach to quantum Brownian motion. Phys A 121:587–616. https://doi.org/10.1016/0378-4371(83)90013-4
19. Slobodeniuk AO, Levkivskyi IP, Sukhorukov EV (2013) Equilibration of quantum Hall edge states by an Ohmic contact. Phys Rev B 88:165307. https://doi.org/10.1103/PhysRevB.88.165307
20. Johnson JB (1928) Thermal agitation of electricity in conductors. Phys Rev 32:97–109. https://doi.org/10.1103/PhysRev.32.97

Chapter 4
Transmission Line—Nonlocal Effects

Abstract This chapter explores the impact of nonlocal interactions on heat transport in quantum Hall edge states using an extended transmission line (TL) model. Building on the previous chapter, where dissipation alone was shown to preserve the quantization of heat flux, we now introduce cross-capacitive coupling and tunnel junctions between Ohmic reservoirs. These nonlocal interactions break the sum rule that ensures heat flux quantization, leading to an enhancement of the heat flux above the expected quantum value. This phenomenon, termed *negative heat drag*, arises from highly correlated charge fluctuations between reservoirs. We analyze both two-node and full TL models, demonstrating how interactions create a second collective mode, which influences heat transport. Additionally, we explore an energy-conserving approach using mesoscopic capacitors, which replicates the observed effects, suggesting that these deviations from quantized heat flux are intrinsic to strongly interacting quantum Hall edges. Our results provide insight into equilibrium heat transport anomalies and offer a framework for designing mesoscopic experiments to probe unconventional heat and charge dynamics in chiral systems.

4.1 Breaking of the Sum Rule

In the previous chapter, we addressed the open question of whether dissipation due to naturally present disorder at the edges can explain the *missing heat paradox*. Surprisingly, we found that even when dissipation of an arbitrary strength is present at equilibrium, the energy flux carried by the edge maintains its universal value. The heat dissipated is precisely offset by the back-action effect of the dissipative degrees of freedom on the chiral edge state, ensuring perfect quantization. This perfect quantization was formulated as a generalized sum rule (3.14), stating that at equilibrium, the heat flux carried by a single edge state always equals a quantum heat flux.

$$\sum_{\substack{\beta=1,2,\\ p=\mathrm{in},\mathrm{oc}}} \left|\mathcal{T}_p^{\alpha\beta}(\omega)\right|^2 G_p = \frac{1}{R_q}, \tag{4.1}$$

F. Stäbler, *Dissipation, Interactions, and Heat Transport in Quantum Hall Devices*, Springer Theses, https://doi.org/10.1007/978-3-031-90671-8_4

In our next attempt, we aim to identify the conditions that lead to the aforementioned sum rule and explore the mechanisms that result in a non-quantum heat flux in a TL system. It becomes evident from the generalized sum rule (4.1) or (3.13) that introducing interactions by coupling the edge states to metallic granules leads to a deviation from the perfect quantization of heat flux [17, 18] when the system is driven out of equilibrium. In this chapter, we focus on a different approach, where edge states can transport a non-quantum amount of heat due to additional interactions. This concept can be understood by examining Fig. 4.1. Unlike our initial approach, which modeled a realistic quantum Hall (QH) edge as a series of bumps and puddles to account for dissipation effects, as depicted in Fig. 3.2, our current model also incorporates the impact of interactions among disorder elements.

Similar phenomena, where interaction between different parts of a system is introduced, have been explored in conventional mesoscopic devices, such as the mutual electron drag effect [19, 20]. In such devices, an active component, driven out of equilibrium, generates a current within a passive segment of the system. There are several known mechanisms for the drag of charge currents, including interactions that involve electron-phonon, electron-photon, and electron-ion couplings [21, 22, 27]. Unlike these mechanisms, there is the thermal drag effect, where a temperature gradient within the system induces a heat current. This heat current is mediated by Coulomb interactions between the active and passive components of the system.

Recently, a similar drag effect was observed in a chiral system composed of two quantum Hall edge states (QH), which interact cross-capacitively through two metallic granules [1]. This interaction leads to a heat drag phenomenon when the active part of the system is driven out of equilibrium. However, this effect is inherently linear; it does not depend on noise rectification through a non-linear element, unlike other comparable drag effects. The concept of cross-capacitive coupling, serving as a form of nonlocal coupling, has also been implemented in nonchiral systems, resulting in a Brownian gyration effect. In such setups, the coupling induces a persistent current, resembling a heat pump or ratchet mechanism, effectively transferring heat from the hotter to the cooler reservoir through the current [28, 29].

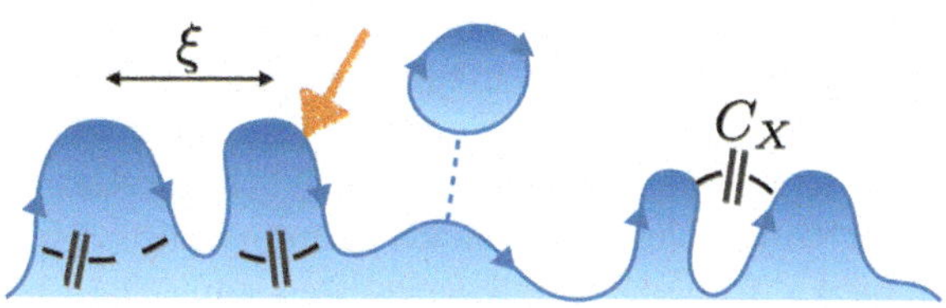

Fig. 4.1 The edge of a Quantum Hall (QH) state may exhibit disorder, leading to the formation of lumps and puddles, similar to those depicted in Fig. 3.2. This disorder is characterized by a length scale, denoted as ξ. In the current chapter, we extend our model to include the interactions between these disordered edge structures. We demonstrate that a local measurement of the effective temperature, as highlighted by the orange arrow, can uncover unexpectedly high effective temperatures, i.e. a non-quantum heat flux, at these disordered regions

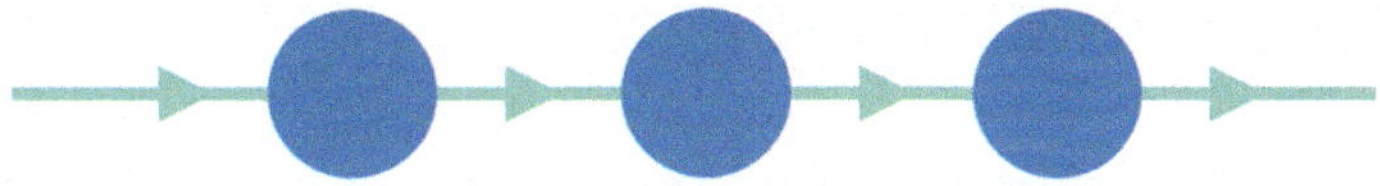

Fig. 4.2 A segment of transmission line. The sum rule states that if the total incoming heat flux is quantum, the outgoing heat flux is also quantum

4.1.1 Backaction Effect of Langevin Sources and Its Exact Cancellation

In Eq. 4.2a, we demonstrated the cancelation between the charge-charge and charge-current correlation functions upon integration over momenta. This was one of the central results of Sect. 3.1.

$$\frac{1}{\tau_C}\int \frac{dk}{2\pi} C_{QQ} = -\int \frac{dk}{2\pi}\left(C_{IQ}+C_{QI}\right), \tag{4.2a}$$

$$C_{XY} = \langle \delta X^{\alpha}(k,\omega)\delta Y^{\alpha}(-k,-\omega)\rangle\,, \tag{4.2b}$$

The precise cancellation of these correlation functions in equilibrium signifies an energy equilibrium; the heat absorbed by a reservoir, represented by $\sim C_{QQ}$, will be subsequently re-emitted as thermal fluctuations, denoted by $\sim C_{QI} + C_{IQ}$. This reciprocal action ensures that the heat transported by the edge state consistently matches an equilibrium heat flux quantum $J \sim C_{II} = J_q$, irrespective of the dissipation strength induced by transverse interactions.

The outcome described above further underscores the inherently local aspect of heat transport and arises directly from it. Consider dividing the transmission line into segments, as illustrated in Fig. 4.2. Regardless of where the transmission line is segmented between reservoirs, if the incoming current is in equilibrium, energy conservation ensures that the outgoing current will also be in equilibrium.

Non-local interactions, that is, interactions between nodes at distinct sites, or the introduction of a second nonchiral channel, significantly alter the dynamics. Energy conservation requires that the total longitudinal cross section carries an equilibrium amount of heat. However, the distribution of energy between the chiral channel and the second channel, introduced by some non-local interaction or diffusion, remains uncertain. This ambiguity already hints at a potential increase or decrease in the heat flux within the chiral channel. To show this effect, a local measurement of the correlation function in the chiral channel, affected by the specific interaction or diffusion, is essential. In this study, we will explore two types of nonlocal interactions or diffusion between the nodes and their respective transmission lines, and calculate the heat transported by the edge state in the presence of these additional correlations among reservoirs.

4.1.2 Cross Capacitive Coupling

Let us consider the simplest nontrivial model that consists of two Ohmic reservoirs coupled by an intermediate edge channel and through direct cross-capacitive coupling of strength C_X between charge fluctuations within each reservoir, as illustrated in Fig. 4.3. The starting point of our analysis is Kirchoff's law and the respective Langevin equations for the outgoing currents of the following form

$$\frac{d}{dt}\Delta Q_1(t) = \delta I_{\mathrm{in}}(t) - \delta I_{\mathrm{out2}}(t), \tag{4.3a}$$

$$\frac{d}{dt}\Delta Q_2(t) = \delta I_{\mathrm{in2}}(t) - \delta I_{\mathrm{out}}(t), \tag{4.3b}$$

$$\delta I_{\mathrm{out2}}(t) = \frac{1}{\tau}\left(\Delta Q_1(t) + \lambda \Delta Q_2(t)\right) + \delta I_{\mathrm{oc2}}(t), \tag{4.3c}$$

$$\delta I_{\mathrm{out}}(t) = \frac{1}{\tau}\left(\Delta Q_2(t) + \lambda \Delta Q_1(t)\right) + \delta I_{\mathrm{oc}}(t), \tag{4.3d}$$

where $\lambda = C/C_X \in 0, 1$, and these equations can be solved for the outgoing currents $\delta \mathbf{I}_{\mathrm{out}} = \left(\delta I \mathrm{out}, \delta I_{\mathrm{out2}}, \delta I_{\mathrm{oc,out}}, \delta I_{\mathrm{oc,out2}}\right)^T$ as functions of the incoming currents and sources $\delta \mathbf{I}_{\mathrm{in}} = (\delta I \mathrm{in}, \delta I_{\mathrm{in2}}, \delta I_{\mathrm{oc}}, \delta I_{\mathrm{oc2}})^T$. This results in a relationship where the Fourier transformation of the currents is expressed in terms of a scattering matrix

$$\delta \mathbf{I}_{\mathrm{out}} = \boldsymbol{\mathcal{M}} \cdot \delta \mathbf{I}_{\mathrm{in}}, \quad \boldsymbol{\mathcal{M}} = \begin{pmatrix} \mathcal{A} & \mathcal{B} & \mathcal{C} & -\mathcal{A} \\ \mathcal{B} & \mathcal{A} & -\mathcal{A} & \mathcal{C} \\ -\mathcal{A} & \mathcal{C} & \mathcal{B} & \mathcal{A} \\ \mathcal{C} & -\mathcal{A} & \mathcal{A} & \mathcal{B} \end{pmatrix}, \tag{4.4}$$

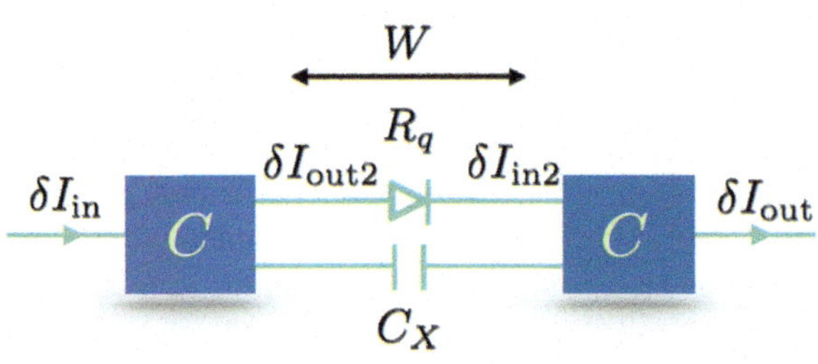

Fig. 4.3 Cross-capacitive coupling of reservoirs. The Hamiltonian density has an additional term of the form $C_X^{-1} Q_j Q_{j'}$, where $j \neq j'$, establishing a direct capacitive coupling between the charge fluctuations of different reservoirs. It is possible to analytically solve the equations of motion for the edge states between the reservoirs, which are influenced by the long-range Coulomb interaction. This solution is expressed in terms of the boundary currents, as shown in the figure. Furthermore, the equations of motion for the bosonic fields or currents are analogous to the Langevin equations, referenced as Eqs. (4.3a)–(4.3d)

with

$$\mathcal{A}(\omega) = \frac{i\lambda\omega\tau_C}{\lambda^2 - (1 - i\omega\tau_C)^2}, \tag{4.5a}$$

$$\mathcal{B}(\omega) = 1 - C, \tag{4.5b}$$

$$C(\omega) = \frac{i\omega\tau_C(1 - i\omega\tau_C)}{\lambda^2 - (1 - i\omega\tau_C)^2}. \tag{4.5c}$$

It can be verified that this scattering matrix is unitary $\boldsymbol{\mathcal{M}}\boldsymbol{\mathcal{M}}^{\dagger} = \mathbb{1}$, highlighting the energy-conserving characteristic of the scattering process prior to integrating the Langevin sources. The scattering matrix $\mathcal{M}$ aligns with the scattering matrix detailed in [1]. Furthermore, we introduce an additional condition that $\delta I_{\text{in}2} = \exp(i\omega W/v_F)\,\delta I_{\text{out}2}$ and proceed to derive the outgoing currents as a function of the incoming boundary current δI_{in}, along with the sources $\delta I_{\text{c,out}}$ and $\delta I_{\text{oc}2}$, respectively. We present the resulting matrix as $(\delta I_{\text{out}}, \delta I_{\text{out}2})^T = \boldsymbol{\mathcal{M}}' (\delta I_{\text{in}}, \delta I_{\text{oc}}, \delta I_{c2})^T$ for the limit $W \to 0$. Note that for finite W this retardation plays the role of an additional cutoff, as it increases the geometric capacitance of the device. Especially in the scenario where $W \to \infty$, the current-current correlation function exhibits rapid oscillations relative to all other energy scales. When averaging over these fast oscillations, the intermediate currents lose their correlations, negating the effect of non-local heat transport, and as a result, all currents carry a quantum of heat flux. However, with a finite W, modulations or resonances in the heat flux are observed. These do not alter the sign of the corrections to the heat flux but naturally emerge due to the modulation of interference between charge fluctuations in the respective reservoirs, facilitated by the non-local interaction.

$$\boldsymbol{\mathcal{M}}' = \begin{pmatrix} \mathcal{D} & \mathcal{F}+\mathcal{G} & \mathcal{E} \\ \mathcal{D}+\mathcal{E} & \mathcal{F} & \mathcal{G} \end{pmatrix}, \tag{4.6a}$$

$$\mathcal{D}(\omega) = \frac{\lambda^2 - 1 + i\lambda\omega\tau_C}{\lambda^2 - i\lambda\omega\tau_C - (1 - i\omega\tau_C)^2}, \tag{4.6b}$$

$$\mathcal{E}(\omega) = \frac{i\omega\tau_C - i\lambda\omega\tau_C}{\lambda^2 - i\lambda\omega\tau_C - (1 - i\omega\tau_C)^2}, \tag{4.6c}$$

$$\mathcal{F}(\omega) = -\frac{i\lambda\omega\tau_C}{\lambda^2 - i\lambda\omega\tau_C - (1 - i\omega\tau_C)^2}, \tag{4.6d}$$

$$\mathcal{G}(\omega) = \frac{i\omega\tau_C(1 - i\omega\tau_C)}{\lambda^2 - i\lambda\omega\tau_C - (1 - i\omega\tau_C)^2}. \tag{4.6e}$$

Note that $\boldsymbol{\mathcal{M}}'$ is not the system's scattering matrix since it incorporates the intermediate current $\delta I_{\text{out}2}$. Therefore, the non-local interaction is responsible for part of the heat transmission.

Next, we proceed under the assumption that the incoming currents and Langevin sources are in equilibrium, sharing the same temperature T.[1] The current-current correlation function, as before (2.22), is described by the equation

$$\left\langle \delta I_p(\omega)\delta I_{p'}(\omega')\right\rangle = 2\pi\delta_{pp'}\delta(\omega+\omega')S_p(\omega) \tag{4.7a}$$

$$S_p(\omega) = \frac{\omega R_q^{-1}}{1-e^{-\omega/T_p}}, \quad p \in \{\text{in}, \text{oc}, \text{oc2}\}. \tag{4.7b}$$

Using Eq. (2.26), we calculate the heat flux transported by the edge states, both after their interaction with and between the reservoirs.

4.1.2.1 Heat Flux Carried by δI_{out}

The heat flux measured after the two nodes is $J_{\text{out}} = R_q/2\left\langle \delta I^2_{\text{out}}(t)\right\rangle$, as indicated in Fig. 4.3. The unitarity of the scattering matrix, which can also be understood as a generalized sum rule, ensures energy conservation throughout the system. This principle is illustrated by the outgoing current-current correlation function (which gives the heat flux after integration):

$$\left\langle \delta I_{\text{out}}(\omega)\delta I_{\text{out}}(\omega')\right\rangle = 2\pi\delta(\omega+\omega')S_{\text{out}}(\omega), \tag{4.8}$$

signifying that the outgoing current fully carries a quantum of heat.

$$S_{\text{out}}(\omega) = S(\omega). \tag{4.9}$$

The noise power $S(\omega)$, as defined in Eq. (4.7a), assuming $T_{\text{oc}\alpha} = T$. By inserting $S(\omega)$ in Eq. (2.26), we indeed retrieve the equilibrium result

$$J_{\text{out}} = J_q. \tag{4.10}$$

4.1.2.2 Heat Flux Carried by δI_{out2}

The heat flux measured after between the two nodes is given by $J_{\text{out2}} = R_q/2\left\langle \delta I^2_{\text{out2}}(t)\right\rangle$, as indicated in Fig. 4.3. The correlation function for $\langle \delta I_{\text{out2}}(\omega)\delta I_{\text{out2}}(-\omega)\rangle$ differs from $S(\omega)$, illustrating that the edge state acquires additional correlations from cross-capacitive interactions. This is captured by the following expressions:

$$\langle \delta I_{\text{out2}}(\omega)\delta I_{\text{out2}}(-\omega)\rangle = 2\pi\delta(\omega+\omega')f(\omega,\lambda)S(\omega), \tag{4.11a}$$

---

[1] Exploring scenarios out of equilibrium, such as connecting the circuit or, subsequently, the transmission line to a hot contact, allows to study features like the relaxation of the incoming distribution function due to dissipation.

$$f(\omega,\lambda) = \frac{\left(\lambda^2-1\right)^2 + \left(2+\lambda^2\right)\omega^2\tau_C^2 + \omega^4\tau_C^4}{\left(\lambda^2-1\right)^2 + (2+\lambda(3\lambda-4))\omega^2\tau_C^2 + \omega^4\tau_C^4}, \tag{4.11b}$$

where $f(\omega, \lambda)$ consistently exceeds 1. In the regime of low temperatures, where $\tau_C T \ll 1, \left(1-\lambda^2\right)$, we can approximate

$$f(\omega,\lambda) = 1 + \frac{2\lambda(2-\lambda)}{\left(1-\lambda^2\right)^2}\omega^2\tau_C^2. \tag{4.12}$$

Upon performing the frequency integral (2.26) to obtain the heat flux we find

$$J_{\text{out2}} = J_q + \frac{2\lambda(2-\lambda)}{\left(1-\lambda^2\right)^2}J_4, \tag{4.13}$$

with $J_4 = \pi^3/30\ \tau_C^2 T^4$. It is important to note that this approximation breaks down as $\lambda \to 1$, nearing the limit of strong interaction. In the case of the strongest possible interaction $\lambda = 1$, we find:

$$f(\omega,1) = \frac{3+\omega^2\tau_C^2}{1+\omega^2\tau_C^2} \approx 3 - 2\omega^2\tau_C^2, \tag{4.14}$$

where we also expanded in $\tau_C T$, which corresponds to the heat Coulomb blockade regime, cf. Chap. 2. This translates to a heat flux in the strongly coupled limit given by

$$J_{\text{out2}} = 3J_q - 2J_4. \tag{4.15}$$

This equation marks the beginning of a novel regime, characterized by an enhancement that begins at the first order $O(T^2)$. By comparing the coefficients in the low temperature expansions, one can determine a critical temperature T_{crit} for a specified λ, or alternatively establish λ_{crit} based on a known T that distinguishes the regime, where $\tau_C T \ll 1, \lambda \approx 0$ and $\tau_C T \ll 1, \lambda \approx 1$.

$$\tau_C T_{\text{crit}} = \frac{\sqrt{\frac{5}{2}}\left(1-\lambda^2\right)}{\pi\sqrt{\lambda\left(\lambda^3-\lambda-2\right)+1}} \tag{4.16}$$

We found that a system with cross-capacitive coupling exhibits negative heat drag, a conclusion explicitly supported by performing the necessary integration in Eq. (2.26) with the application of Eq. (4.11a) in the limit previously discussed. In the leading order and in the strong coupling limit $\lambda = 1$, the heat flux carried by the intermediate chiral edge state δI_{out2} is found to be three times greater than the equilibrium flux J_q, equivalent to an elevation in the effective temperature by a factor of $T_{\text{eff}} \approx \sqrt{3}T$, where T denotes the temperature associated with the injected equilibrium heat flux.

In summary, at low temperatures $\tau_C T$, the heat flux J_{out2} has as a T^4 correction in the case of weak interactions $\lambda \approx 0$ and as a T^2 correction for strong interactions $\lambda \approx 1$, with a positive sign[2] regardless of the value of λ. This amplification is attributed to additional correlations arising from out-of-phase fluctuations of charge across the two reservoirs, leading to a correlated state within the chiral channel. The notion of this correlated state and its implications will be elaborated on in Chap. 5.

4.1.3 Cross Capacitive Coupling—Transmission Line

The TL model represents more than a mere expansion of the two-reservoir framework; it introduces novel physical phenomena, specifically the emergence of a second collective mode, i.e. the out-of-phase fluctuations in the two-node case. This development has significant implications for the dynamics of heat transport. We consider a TL composed of reservoirs interconnected through cross-capacitive interactions, as depicted in Fig. 4.4.

In this model, each chiral edge channel is accompanied by a second, nonlocal channel at every cross section of the infinite line. The equations of motion for this scenario are provided as follows:

$$\frac{d}{dt}Q(j,t) = \delta I_{\text{in}}(j,t) - \delta I_{\text{out}}(j,t), \tag{4.17a}$$

$$\delta I_{\text{out}}(j,t) = \frac{1}{\tau_C}Q(j,t) + \delta I_{\text{oc}}(j,t), \tag{4.17b}$$

$$Q(j,t)(t) = Q(j,t)(t) + \lambda\left(Q(j-1,t) + Q(j+1,t)\right), \tag{4.17c}$$

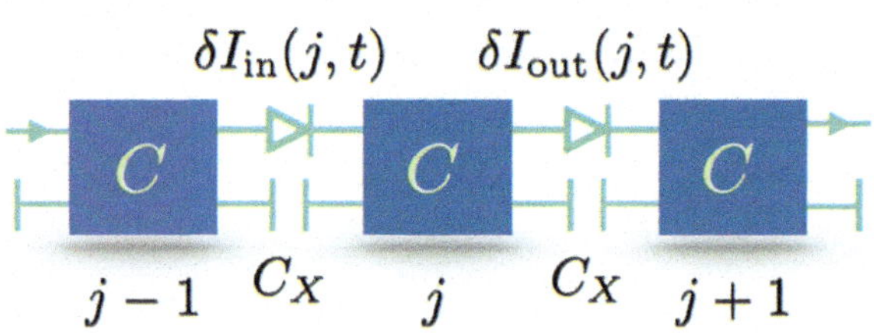

Fig. 4.4 Transmission line with cross-capacitive coupling. This model extends the two-node framework by incorporating nearest-neighbor capacitive interactions between reservoirs. As with the earlier approach, the equations of motion can be derived either by directly solving the Hamiltonian equations of motion or by addressing the corresponding Langevin equations

[2] This was checked by a numerical integration of the exact expression.

where the first equation ensures charge conservation at each node, similar to Eq. (3.2), while the second equation modifies the Langevin framework to incorporate non-local interactions, parameterized by $\lambda = C/C_X$, similar to Eq. (3.2b). These equations are then promoted to transmission line equations by enforcing continuity,[3] such that $\delta I_{\text{in}}(j,\omega) = \delta I_{\text{out}}(j-1,\omega)$. Similarly to before, noting that dissipation attenuates the collective mode, we consider a system with a large number of nodes and impose periodic boundary conditions. This setup then allows the use of discrete Fourier transformations.

$$X(j,t) = \sum_{k=0}^{N-1} \int \frac{d\omega}{2\pi} e^{i\frac{2\pi k}{N}j - i\omega t} X(k,\omega), \tag{4.18a}$$

$$X(k,\omega) = \frac{1}{N} \sum_{j=0}^{N-1} \int dt e^{-i\frac{2\pi k}{N}j + i\omega t} X(j,t), \tag{4.18b}$$

In the limit as the number of nodes N approaches infinity, the discrete sum over the wave numbers becomes a continuous integral:

$$\lim_{N\to\infty} \frac{1}{N} \sum_{k=0}^{N-1} \to \int_{-\pi/\xi}^{\pi/\xi} \frac{dk}{2\pi}, \tag{4.19}$$

where ξ denotes the spacing between the nodes. To calculate the heat flux transported by the edge state between reservoir j and $j+1$, we follow a methodology similar to that of the preceding section: (i) Do a Fourier transformation to solve for $\delta I_{\text{out}}(k,\omega)$ using Eqs. (4.17a)–(4.17c); (ii) Determine the correlation function $\langle \delta I_{\text{out}}(k,\omega)\delta I_{\text{out}}(k',\omega')\rangle = 2\pi\delta(\omega+\omega')\delta(k+k')S_{\text{out}}(k,\omega)$ and execute an integration over both momenta and frequencies; (iii) Address the momentum integral by converting it to a unit circle contour with the transformation $z \to \exp(ik\xi)$, and then identify the residues within the contour.The momentum dependent noise power for the outgoing current (i.e. in between the nodes) is given by

$$S_{\text{out}}(k,\omega) = \frac{\omega^2\, S(\omega)}{(\omega - \varepsilon(k))\,(\omega + \varepsilon(-k))}, \tag{4.20}$$

where $S(\omega)$ is given by Eq. 4.7a assuming all temperatures are equal and the spectrum of the collective mode, $\varepsilon(k)$, is determined by

$$\varepsilon(k) = \frac{i}{\tau_C}\left(e^{-ik\xi} - 1\right)(1 + 2\lambda\cos(k\xi)). \tag{4.21}$$

The eigenvalues of the capacitance matrix, described by

---

[3] Adding retardation in form of a phase factor between the nodes does not change the results qualitatively and we thus omit it.

$$C_{nm} = C^{-1}\left(\delta_{n,m} + \lambda\left(\delta_{n,m-1} + \delta_{n,m+1}\right)\right), \tag{4.22}$$

are $C^{-1}(1 + 2\lambda\cos(k\xi))$. These eigenvalues must be positive definite, constraining λ to the range 0, 1/2. The momentum-integrated outgoing noise power function can be represented through a contour integral as

$$\tilde{S}_{\text{out}}(\omega) = \oint_C \frac{dz}{2\pi i} \frac{\omega^2 S(\omega)}{(z\omega - z\tilde{\varepsilon}(z))(\omega + \tilde{\varepsilon}(1/z))}, \tag{4.23}$$

where $\tilde{\varepsilon}(z) = \frac{i}{\tau_C}\left(\frac{1}{z} - 1\right)\left(1 + \lambda\left(z + \frac{1}{z}\right)\right)$. The noise power, $\tilde{S}_{\text{out}}(\omega) = \int dk/(2\pi)\, S_{\text{out}}(k, \omega)$, is defined as the integral over momentum of $S_{\text{out}}(k, \omega)$. Generally, this integral reveals six poles with a complex structure. We present the solution to the integral for both non-interacting ($\lambda \to 0$) and strongly interacting ($\lambda \to 1/2$) limits, allowing for the calculation of low-temperature deviations from these benchmarks and identifying the crossover energy scale between the two cases. Subsequently, we will analyze the heat flux $J_{\text{out}} = R_q/(4\pi)\int d\omega\, \tilde{S}_{\text{out}}(\omega) - \text{vac.}$ for weak and strong coupling.

4.1.3.1 Non Interacting Limit $\lambda = 0$

In this scenario, the integral simplifies to the form presented in Eq. (3.9), as discussed in Sect. 3.1. Within the contour, there is a single pole at $z_1 = (1 - i\omega\tau_C)^{-1}$, while another pole, $z_2 = 1 + i\omega\tau_C$, lies outside the contour. Focusing on the inside pole, z_1, we find that

$$\tilde{S}_{\text{out}}(\omega) = S(\omega),$$

results in a quantum of heat flux. This outcome is attributed to the localized nature of the interaction and the unitarity of the scattering matrix, echoing findings from the previous section.

4.1.3.2 Weakly Interacting Limit $\lambda \to 0$

Through a perturbative expansion, considering $\tau_C T \ll 1, (1 - 2\lambda)$ we find a correction to the heat flux is given by:

$$J_{\text{out}} = J_q + 8\pi J_4 \mathcal{J}(\lambda), \tag{4.24a}$$

$$\mathcal{J}(\lambda) = \frac{\lambda}{(1 + 2\lambda)^3} + \frac{\lambda(1 - \lambda)}{(1 - 2\lambda)^{3/2}(1 + 2\lambda)^{5/2}}, \tag{4.24b}$$

with $J_4 = \pi^3/30\, \tau_C^2 T^4$.

4.1.3.3 Strongly Interacting Limit $\lambda \to \frac{1}{2}$

In the regime of strong interactions, the poles for the integral specified in Eq. (4.23) are derived from solving two cubic equations:

$$1 - z^2 - z^3 + z(1 + 2i\tau_C\omega) \stackrel{!}{=} 0, \tag{4.25a}$$

$$1 + z - z^3 - z^2(1 - 2i\tau_C\omega) \stackrel{!}{=} 0. \tag{4.25b}$$

It can be seen that one pole consistently lies outside the contour, while another remains inside for all values of ω. Additionally, two poles are inside the contour exclusively for $\omega > 0$, and the remaining two poles are found inside only for $\omega < 0$. A detailed analysis of the pole structure is presented in Appendix B.1. Selecting the poles within the contour reveals that, at low temperatures, the enhancement to the heat flux quantum can be expressed as:

$$J_{\text{out}} = J_q + \frac{3}{8}\sqrt{\frac{\pi\tau_C T^3}{2}}\zeta\left(\frac{5}{2}\right), \tag{4.26}$$

where $\zeta(s) = \sum_{n=1}^{\infty} n^{-s}$ denotes the Riemann-Zeta function. In contrast to the two node case we find yet another enhancement of the heat flux that has a correction that scales as $T^{3/2}$.

4.1.3.4 Crossover Between the Two Limits

Upon analyzing $\mathcal{J}(\lambda)$, we see that it diverges as $\lambda \to 1/2$. This signifies the breakdown of the perturbative expansion since the difference $1 - 2\lambda$ may become smaller as the energy scale set by temperature. This is the point where a second soft mode emerges. This can be seen from the pole structure of the propagator, the inverse Fourier transformation of $(-i\omega - \varepsilon(k))^{-1}$, with the spectrum from Eq. (4.21), see also Fig. 4.5 for the pole structure of the propagator.

In the strongly interacting limit there will be an additional pole at $\omega = 0,\ k = \pi$ in addition to the pole at $\omega = 0,\ k = 0$. We can estimate the crossover from the single-mode solution to the two-mode solution by comparing the coefficients of Eqs. (4.26) and (4.24a). The threshold for this transition, whether it is the critical temperature T_{crit} for a specified λ or the critical λ_{crit} for a given temperature, can be quantified by equating the coefficients from Eqs. (4.26) and (4.24a), resulting in:

$$\tau_C T_{\text{crit}} = \frac{\left(\frac{45}{2}\zeta\left(\frac{5}{2}\right)\right)^{2/3}}{8\pi^{7/3}}\left(\frac{1}{\mathcal{J}(\lambda)}\right)^{2/3}. \tag{4.27}$$

Note that the non-local effect due to cross-capacitive coupling is dynamic. In the limit where the collective mode is not excited ($\tau_C \to 0$), the excess heat flux vanishes;

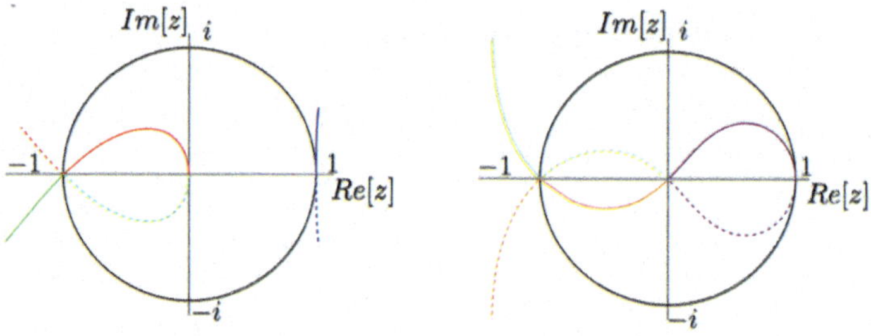

(a) (b)

Fig. 4.5 Poles of the heat integral in Eq. (4.23). The dashed lines indicate $\omega < 0$ and solid lines $\omega > 0$. The blue and purple poles are always present. In the non-interacting limit $\lambda \to 0$ we only pickup the purple pole. The other poles around $z = -1 \;\to\; k = \pi$ correspond to the second soft mode, the out-of-phase fluctuations of the collective mode. For finite λ these poles move towards the center of the contour, i.e. the momentum obtains a finite imaginary part. The excitation becomes gaped and we can rely on our low energy expansion, where the smallest energy scale is set by temperature, given by Eq. 4.24a

this is easily seen from Eq. (4.11b). The nonlocal heat transport can be understood as arising from out-of-phase fluctuations ("$k = \pi$"), leading to heat enhancement in the edge state. The limits of strong interactions yield different outcomes for the TL compared to the two-node model. This distinction is evident because, for $\lambda_{2\text{node}} = 1$, the two-node model simplifies to a single node with charging energy $\frac{1}{2C}(Q_1 + Q_2)^2$, while this simplification does not apply to the transmission line at $\lambda_{\text{TL}} \to 1/2$. The system's Hamiltonian includes only the interaction of neighboring charges $\frac{1}{2C}\sum_j \left(Q_j^2 + Q_j Q_{j+1}\right)$, not equivalent to $\frac{1}{2C}\left(\sum_j Q_j\right)^2$.

4.1.3.5 Work Done by One Reservoir on the Neighboring Reservoirs

To find the part of heat passed by the non-local interaction, envision enclosing a single node of the transmission line depicted in Fig. 4.4 within a hypothetical box of length L. This box is such that the bosonic current density $I(x, t) = -\frac{e}{2\pi}\partial_t \phi(x, t)$ at its borders matches $I(0, t) = I_{\text{in}}(j, t)$ and $I(L, t) = I_{\text{out}}(j, t)$. Subsequently, write a continuity equation for the Hamiltonian density within this segment of the transmission line, $\partial_t \left\langle \hat{\mathcal{H}}_j \right\rangle + \partial_x J_j = 0$, where j signifies the position of the node. In the context of the reservoir model, the limit $L \to \infty$ is ultimately applied. For further elaboration, see Appendix 4.1.6. The Hamiltonian is given by:

$$\mathcal{H}_j = \frac{v_F}{4\pi}\int_0^L dx \left(\partial_x \phi_j(x, t)\right)^2 + \frac{Q_j^2(t)}{2C} + \frac{Q_j(t) Q_{j-1}(t)}{C_X}, \tag{4.28}$$

with $Q_j(t) = \frac{e}{2\pi}\int_0^L dx\; \partial_x \phi_j(x, t)$ as the integrated charge density for the node.

The total energy current density in the box $J_j = -\partial_t \langle \mathcal{H}_j \rangle = 0$ is zero since the total energy is conserved, but we are able to express everything in terms of incoming and outgoing heat fluxes plus some additional term.

$$J_j = \frac{R_q}{2} \left[\langle \delta I^2(L,t) \rangle - \langle \delta I^2(0,t) \rangle \right] + \lambda \left\langle Q_j(t) \partial_t Q_{j-1}(t) - Q_j(t) \partial_t Q_{j+1}(t) \right\rangle . \quad (4.29)$$

The first two terms correspond to the heat flux outgoing $J_{\text{out}} \sim \langle \delta I^2(L,t) \rangle$ and incident $J_{\text{in}} \sim \langle \delta I^2(0,t) \rangle$ to the reservoir.[4] In the absence of the cross capacitive coupling, these two terms cancel exactly, due to the cancellation of Joule heating and the back-action effect of the Langevin sources discussed in Sect. 4.1.1.

In the context of cross capacitive interaction, the additional terms outlined in Eq. (4.29) are the reciprocal work exchanges between adjacent reservoirs. Specifically, within a two-node model, we see that

$$-i\omega\lambda \, \langle Q_1(\omega) Q_2(-\omega) \rangle - \text{h.c.} \sim 1 - f(\omega, \lambda), \quad (4.30)$$

which precisely offsets the surplus heat within the chiral channel. Consequently, each cross-section effectively transmits only a single quantum of heat, aligning with thermodynamic predictions. Nevertheless, local measurements might detect an unusual amount of heat in the chiral channel.

It is noteworthy that Eq. (4.28) is the identical Hamiltonian discussed in [1], where the phenomenon of heat drag in chiral systems was explored. Hence, our examination extends to this scenario as well, enabling the determination of heat transfer through cross capacitive coupling across all examined regimes in [1].

4.1.4 *Tunnel Coupling*

The second example of non-local interaction that we examine involves the direct tunneling between reservoirs via a quantum point contact (QPC), as depicted in Fig. 4.6. The equations of motion are given by the following set of equations:

$$\frac{d}{dt} \Delta Q_1(t) = \delta I_{\text{in}}(t) - \delta I_{\text{out2}}(t) - \delta I_{\Omega R}(t) + \delta I_{R\Omega}(t), \quad (4.31\text{a})$$

$$\frac{d}{dt} \Delta Q_2(t) = \delta I_{\text{in2}}(t) - \delta I_{\text{out}}(t) - \delta I_{\Omega L}(t) + \delta I_{L\Omega}(t), \quad (4.31\text{b})$$

where these two equations are Kirchhoff's law, ensuring charge conservation for the two reservoirs. Incoming currents $\delta I_{L\Omega}$ and $\delta I_{R\Omega}$, passing through the QPC, are characterized by an energy-independent scattering matrix ($\mathcal{P} \in 0, 1$) at the QPC.

[4] Assuming subtraction of vacuum contributions everywhere.

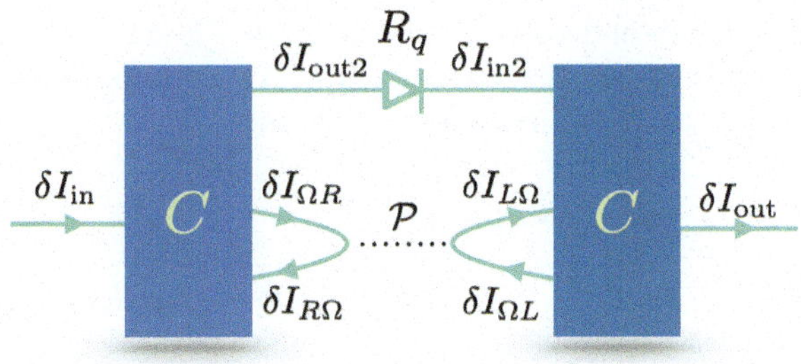

Fig. 4.6 Direct tunnel coupling of the reservoirs. We consider QPCs in the regime where the transmission probability is energy independent. In this context, bosonic current tunneling exhibits non-unitary behavior, yet the partitioning noise adapts to ensure energy equilibrium is maintained between the incoming and outgoing heat fluxes at the QPC

Note that the assumption of energy-independent transmission applies primarily when considering the tunneling of free fermions. In the leading approximation of tunneling, collective modes and source contributions are considered uncorrelated, justifying this simplification. Our analysis is valid for $\mathcal{P}$ values close to 0 or 1, and a further examination is necessary for intermediate values $\mathcal{P}$, which will be addressed in future work, also see Chap. 5. The currents crossing the quantum point contact (QPC) are described by:

$$\delta I_{R\Omega}(t) = \mathcal{P}\delta I_{\Omega L}(t) + (1-\mathcal{P})\delta I_{\Omega R}(t) + \delta I_{\text{QPC}}(t), \tag{4.32a}$$

$$\delta I_{L\Omega}(t) = \mathcal{P}\delta I_{\Omega R}(t) + (1-\mathcal{P})\delta I_{\Omega L}(t) - \delta I_{\text{QPC}}(t), \tag{4.32b}$$

where the process is inherently non-unitary, i.e. not conserving the energy. However, the QPC introduces partitioning noise δI_{QPC}, with its noise power tailored to ensure energy conservation throughout the QPC. Consequently, the energy fluxes satisfy $J_{R\Omega} + J_{L\Omega} = J_{\Omega R} + J_{\Omega L}$, thus maintaining the energy balance. For an in-depth discussion of this topic, see the work by [17] on electrical noise and energy conservation at the QPC.

The outgoing currents from the reservoirs are described by Langevin equations as follows:

$$\delta I_{\Omega R} = \frac{1}{\tau_C} Q_1(t) + \delta I_{\text{oc},R}(t), \tag{4.33a}$$

$$\delta I_{\Omega L} = \frac{1}{\tau_C} Q_2(t) + \delta I_{\text{oc},L}(t), \tag{4.33b}$$

$$\delta I_{\text{out2}}(t) = \frac{1}{\tau_C} Q_1(t) + \delta I_{\text{oc},2}(t), \tag{4.33c}$$

$$\delta I_{\text{out}}(t) = \frac{1}{\tau_C} Q_2(t) + \delta I_{\text{oc}}(t). \tag{4.33d}$$

We again assume that there is no retardation in the intermediate current, setting $\delta I_{\text{in2}}(t) = \delta I_{\text{oc,out2}}(t)$. The equations can be solved by a Fourier transform as before and we again assume that the temperatures of all the reservoirs match the boundary current temperature to the left and have a noise power given by $S(\omega)$ defined in (4.7a). Each cross section carries an equilibrium heat amount, which can be seen by the fact that $J_{\text{out2}} + J_{\Omega R} - J_{R\Omega} = J_q$ or $J_{\text{out2}} + J_{\Omega L} - J_{L\Omega} = J_q$, according to the relation described in Eq. (2.26).

However, each individual channel, between the reservoirs, carries a non-quantum amount of heat. The noise power of the outgoing intermediate current is given by

$$\langle \delta I_{\text{out2}}(\omega)\delta I_{\text{out2}}(\omega')\rangle = 2\pi\delta(\omega+\omega')\,(1+g(\omega,\mathcal{P}))\,S(\omega), \tag{4.34a}$$

$$g(\omega,\mathcal{P}) = \frac{2\mathcal{P}\left(1+\mathcal{P}-\omega^2\tau_C^2\right)}{(1+\mathcal{P})^2+2(1+\mathcal{P})(1+2\mathcal{P})\omega^2\tau_C^2+\omega^4\tau_C^4}. \tag{4.34b}$$

Similarly to the cross-capacitive coupling, we can perform a perturbative expansion for $\tau_C T \ll 1$ and find

$$g(\omega,\mathcal{P}) \approx \frac{2\mathcal{P}}{1+\mathcal{P}} - \frac{2\mathcal{P}(3+4\mathcal{P})\omega^2\tau_C^2}{(1+\mathcal{P})^2}. \tag{4.35}$$

The correction thus starts already at the constant order and hence represents a direct correction to the value of the heat flux quantum. The heat carried by this channel is given by

$$J_{\text{out2}} = \left(1+\frac{2\mathcal{P}}{1+\mathcal{P}}\right)J_q - \frac{2\mathcal{P}(3+4\mathcal{P})}{(1+\mathcal{P})^2}J_4, \tag{4.36}$$

where $J_4 = \frac{\pi^3}{30}\tau_C^2 T^4$. Note that in contrast to the cross-capacitive coupling Eq. (4.11b), this is not a dynamical effect. The term $2\mathcal{P}/(1+\mathcal{P})J_q$ survives in the limit where all dynamics are suppressed $\tau_C \to 0$. This represents a static contribution, a frequency independent neutral excitation of electron-hole pairs, which create additional correlations on the chiral channel.

4.1.5 Tunnel Coupling—Transmission Line

Similarly to before, we promote the two-reservoir model to a transmission line. The modified equations of motion have the following form.

$$\frac{d}{dt}Q(j,t) = \delta I_{\text{in}}(j,t) - \delta I_{\text{out}}(j,t) + \delta I_{\text{QPC}}(j,t), \tag{4.37}$$

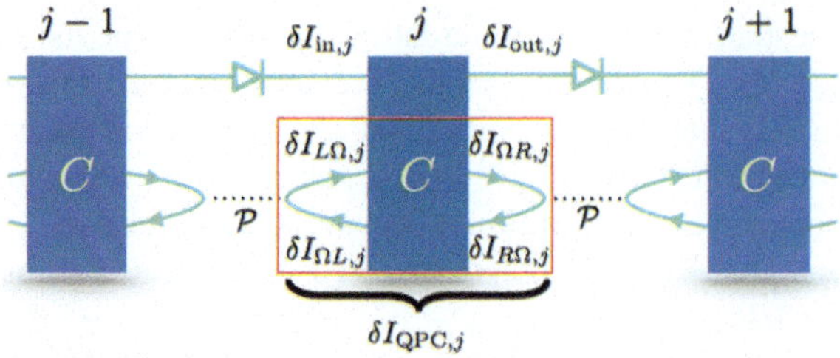

Fig. 4.7 Transmission line with tunnel coupling. We extend our two node model by introducing nearest neighbor tunneling between reservoirs in the form of a QPC. Similar to before the equations of motion can be either found from solving the Hamiltonian equations of motion directly or by solving the corresponding Langevin equations

where n labels the position of the node in the line and $\delta I_{\text{QPC}}(j,t) = \delta I_{L\Omega}(j,t) - \delta I_{\Omega L}(j,t) + \delta I_{R\Omega}(j,t) - \delta I_{\Omega R}(j,t)$, is the total net current of the incoming and outgoing from the left and right QPC; see Fig. 4.7.

To obtain the transmission line equations, we demand $\delta I_{\text{in}}(j,\omega) = \delta I_{\text{out}}(j-1,\omega)$. Since dissipation leads to a decay of the collective mode, we will consider a large number of nodes and assume periodic boundary conditions, which allows us to use discrete Fourier transformations, Eqs. (4.18a) and (4.18b). The solution of the heat flux follows the same steps as already presented in Sect. 4.1.3. The outgoing current $\delta I_{\text{out}}(j,t)$ satisfies a normal Langevin equation. However, the QPC current satisfies a special Langevin equation of the following form.

$$\delta I_{\text{QPC}}(k,\omega) = -\frac{2\mathcal{P}}{\tau_C}\left(1-\cos(k\xi)\right) Q(k,\omega) + \delta I_{\text{oc,QPC}}(k,\omega), \tag{4.38}$$

with a noisepower $S_{\text{QPC}} = (2R_q/R_{\text{QPC}})S(\omega)$ with a momentum depended conductance $R_{\text{QPC}}^{-1} = 2\mathcal{P}R_q^{-1}\left(1-\cos(k\xi)\right)$. Note the difference of factor two due to non-chirality of the Langevin source. The sign in Eq. (4.38) is due to the convention of the definition of current, however R_{QPC}^{-1} can be understood as the eigenvalues of the capacitance matrix of the system and are defined to be positive definite. The spectrum of the collective mode is given by

$$\varepsilon(k) = \frac{i}{\tau_C}\left(e^{-ik\xi}-1\right) - \frac{2i\mathcal{P}}{\tau_C}\left(1-\cos(k\xi)\right). \tag{4.39}$$

The outgoing noise power at node j in the chiral channel is given by

$$\left\langle \delta I_{\text{out}}(k,\omega)\delta I_{\text{out}}(k',\omega')\right\rangle = 2\pi\delta(k+k')\delta(\omega+\omega')S_{\text{out}}(\omega), \tag{4.40}$$

where

$$S_{\text{out}}(\omega) = \frac{S_{\text{QPC}}(\omega)}{\tau_C^2(\omega + \epsilon(-k))(\omega - \epsilon(k))} + \frac{S(\omega)\left(1 + e^{ik\xi}(i\tau_C(-\omega+\epsilon(k))-1)\right)(\text{h.c.})}{\tau_C^2(\omega + \epsilon(-k))(\omega - \epsilon(k))}. \tag{4.41}$$

We map this onto the unit circle contour by doing the transformation $z \to e^{ik\xi}$ and pick up the poles inside of the contour. We find in total 6 poles, but in contrast to the capacitive coupling they stay inside or outside independently of the sign of ω. For the momentum integrated noise power $\tilde{S}_{\text{out}}(\omega)$ at low temperatures $\tau_C T \ll 1$ we find

$$\tilde{S}_{\text{out}}(\omega) \approx 1 + \frac{2\mathcal{P}}{1+\mathcal{P}} - 6\mathcal{P}(1+2\mathcal{P})\,\omega^2\tau_C^2. \tag{4.42}$$

This yields a heat flux, integrating over momenta as before

$$J_{\text{out}} = \left(1 + \frac{2\mathcal{P}}{1+\mathcal{P}}\right) J_q - 6\mathcal{P}(1+2\mathcal{P})\, J_4, \tag{4.43}$$

where $J_4 = \frac{\pi^3}{30}\tau_C^2 T^4$. The heat flux is qualitatively similar to the one obtained for the two node case. Note also here the emergence of the direct enhancement of the heat flux quantum in the chiral channel. The heat carried by the QPC current is negative and exactly compensates the excess heat.

4.1.6 Mesoscopic Capacitors—Energy Conserving Approach

One central assumption in this chapter is that all dissipative degrees of freedom lie in the OCs, which typically represent a metallic system with an infinitely small level spacing. This condition can be relaxed by replacing the ohmic reservoirs with mesoscopic capacitors [23, 24], a fully Hamiltonian energy conserving system without dissipation. A mesoscopic capacitor is a loop of an edge state, which interacts via long-range Coulomb interactions with itself; see Fig. 4.8. The advantage of this is twofold. On the one hand, this mesoscopic capacitor is an easily realizable mesoscopic device; an alternative building block to an OC. On the other hand, this mesoscopic capacitor might be a more realistic approximation of the shape and form of disorder and defects inherent to QH edges. Our original motivation was to consider an effective model that intrinsically includes dissipation. However, the assumption of modeling this as an OC remained an open question. As we show, many of our results for the reservoirs also qualitatively apply to the energy conserving approach and might be an inherent property of QH edges.

In Appendix B.2 we briefly revisit the cases of cross-capacitive coupling with mesoscopic capacitors instead of OCs. Note the difference. The former does not

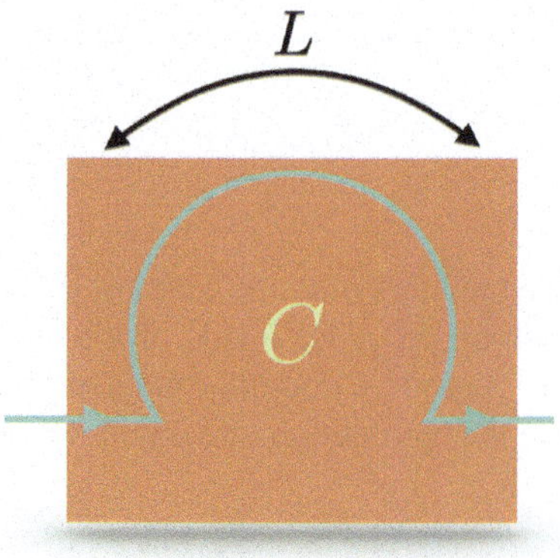

Fig. 4.8 Mesocopic capacitor. A loop of edges state subject to Coulomb self interaction. The Hamiltonian inside of the loop acquires the additional interaction term $Q^2/2C$. Where Q is the total charge inside of the loop

rely on the Langevin formalism, but on scattering theory, and is a fully Hamiltonian system without any dissipation. As an example, we considered two mesoscopic capacitors subject to cross-capacitive interaction, but the special role of the mesoscopic capacitors can be directly generalized to tunnel coupling and the respective transmission lines. The main results are that a large capacitor $L \to \infty$ in equilibrium is equivalent to the reservoir model considered in this paper. The reason for this is that correlations between the currents propagating inside of the capacitor are lost, in a similar way as current is dissipated in the OC, which leads to the same physical situation and prediction for the correction of heat flux due to non-local interactions. Our analysis holds if the following separation of the energy scale is fulfilled $v_F/L \ll T \ll E_c$, where the temperature T is larger than the level spacing, but smaller than the capacitor charging energy.

For a finite length L the result remains qualitatively the same as for the reservoirs, however the correlation function develops additional resonances due to interference effects between the charge fluctuations of the reservoirs.

4.1.7 Concluding Remarks

In this part of the project, we model chiral systems with effective instrinsic dissipation in the form of ohmic reservoirs. We rely on a combination of scattering theory and Langevin equations and have full access to equilibrium and non-equilibrium situations. In thermal equilibrium, there is a cancellation between the Joule heating of the reservoirs and a back-action effect of the Langevin sources. This cancellation is the reason for a robust quantization of heat carried by the edge state in the presence of dissipation and in the absence of chirality breaking non-local interactions or diffusion.

We have shown that non-local interactions or diffusive currents between the reservoirs in a chiral system create highly correlated and nontrivial states, leading to a negative heat drag effect; an enhancement of the heat flux above the quantized value of the heat flux quantum $J_q = \pi/12\ T^2$ and thus a violation of the aforementioned energy balance in the chiral channel. This nonlocal passage of energy is in full agreement with the second law of thermodynamics, since every cross section of the system only carries an equilibrium amount of heat. This enhancement can be attributed to out of phase fluctuations for cross capacitively coupled reservoirs and to an enhancement through the Coulomb blockade effect in the case of a tunnel coupling.

We have shown that neither dissipation nor the specific choice of our model for a reservoir is the cause of this phenomenon. The same effect can be found by replacing the reservoirs with mesoscopic capacitors, a fully energy-conserving system, consisting of a loop of the edge state subject to Coulomb self-interaction. For a large mesoscopic capacitors the predictions are the same as for the reservoirs. For small mesoscopic capacitors the current-current correlation function develops resonances, but the effect remains qualitatively the same.

Our results show the non-universality involved in heat transport experiments already in equilibrium, and this paper can be understood as a toolbox to construct and detect these highly correlated and non-trivial states in mesoscopic experiments.

To proceed further, one needs to look into the physics of probing the correlated states, which is the starting point of paper III. The states are highly correlated, and thus in the tunneling Hamiltonian approach one needs to define an electronic operator which is not just the bosonized vertex operator but goes beyond the bosonization formalism due to correlations through the collective mode if a tunnel probe is connected. In paper III we answer the question if one has experimental access to the charge of the excitations, e.g. by Aharonov-Bohm type experiments [25], to the quantum statistics, e.g. by nonequilibrium measurements [26] or to the correlation function directly by tunneling to a QD [1]. Finally, the formalism developed in this project can be easily applied to construct mesoscopic devices showing anomalous heat transport, unusual statistics, and more interesting phenomena.

References

1. le Sueur H, Altimiras C, Gennser U, Cavanna A, Mailly D, Pierre F (2010) Energy relaxation in the integer quantum Hall regime. Phys Rev Lett 105:056803. https://doi.org/10.1103/PhysRevLett.105.056803
2. Granger G, Eisenstein JP, Reno JL (2009) Observation of chiral heat transport in the quantum Hall regime. Phys Rev Lett 102:086803. https://doi.org/10.1103/PhysRevLett.102.086803
3. Venkatachalam V, Hart S, Pfeiffer L, West K, Yacoby A (2012) Local thermometry of neutral modes on the quantum Hall edge. Nat Phys 8:676–681. https://doi.org/10.1038/nphys2384
4. Goremykina A, Borin A, Sukhorukov E (2019) Heat current in a dissipative quantum Hall edge. arXiv:1908.01213
5. Aleiner IL, Glazman LI (1994) Novel edge excitations of two-dimensional electron liquid in a magnetic field. Phys Rev Lett 72:2935–2938. https://doi.org/10.1103/PhysRevLett.72.2935
6. Chklovskii DB, Shklovskii BI, Glazman LI (1992) Electrostatics of edge channels. Phys Rev B 46:4026–4034. https://doi.org/10.1103/PhysRevB.46.4026
7. de Chamon CC, Wen XG (1994) Sharp and smooth boundaries of quantum Hall liquids. Phys. Rev. B 49:8227–8241. https://doi.org/10.1103/PhysRevB.49.8227

8. Rodriguez RH, Parmentier FD, Ferraro D, Roulleau P, Gennser U, Cavanna A, Sassetti M, Portier F, Mailly D, Roche P (2020) Relaxation and revival of quasiparticles injected in an interacting quantum Hall liquid. Nat Commun 11:2426. https://doi.org/10.1038/s41467-020-16331-4
9. Lunde AM, Nigg SE (2016) Statistical theory of relaxation of high-energy electrons in quantum Hall edge states. Phys Rev B 94:045409. https://doi.org/10.1103/PhysRevB.94.045409
10. Fischer SG, Park J, Meir Y, Gefen Y (2019) Interaction-induced charge transfer in a mesoscopic electron spectrometer. Phys Rev B 100:195411. https://doi.org/10.1103/PhysRevB.100.195411
11. Krähenmann T, Fischer SG, Röösli M, Ihn T, Reichl C, Wegscheider W, Ensslin K, Gefen Y, Meir Y (2019) Auger-spectroscopy in quantum Hall edge channels and the missing energy problem. Nat Commun 10:3915. https://doi.org/10.1038/s41467-019-11888-1
12. Shamim S, Shekhar P, Beugeling W, Böttcher J, Budewitz A, Mayer J-B, Lunczer L, Hankiewicz EM, Buhmann H, Molenkamp LW (2022) Counterpropagating topological and quantum Hall edge channels. Nat Commun 13:2682. https://doi.org/10.1038/s41467-022-29815-2
13. Sukhorukov EV (2016) Scattering theory approach to bosonization of non-equilibrium mesoscopic systems. Physica E 77:191–198. https://doi.org/10.1016/j.physe.2015.11.018
14. Feynman RP, Vernon FL (1963) The theory of a general quantum system interacting with a linear dissipative system. Ann Phys 24:118–173. https://doi.org/10.1016/0003-4916(63)90068-X
15. Caldeira AO, Leggett AJ (1983) Path integral approach to quantum Brownian motion. Phys A 121:587–616. https://doi.org/10.1016/0378-4371(83)90013-4
16. Slobodeniuk AO, Levkivskyi IP, Sukhorukov EV (2013) Equilibration of quantum Hall edge states by an Ohmic contact. Phys Rev B 88:165307. https://doi.org/10.1103/PhysRevB.88.165307
17. Sivre E, Duprez H, Anthore A, Aassime A, Parmentier FD, Cavanna A, Ouerghi A, Gennser U, Pierre F (2019) Electronic heat flow and thermal shot noise in quantum circuits. Nat Commun 10:5638. https://doi.org/10.1038/s41467-019-13566-8
18. Duprez H, Pierre F, Sivre E, Aassime A, Parmentier FD, Cavanna A, Ouerghi A, Gennser U, Safi I, Mora C, Anthore A (2021) Dynamical Coulomb blockade under a temperature bias. Phys Rev Res 3:023122. https://doi.org/10.1103/PhysRevResearch.3.023122
19. Narozhny BN, Levchenko A (2016) Coulomb drag. Rev Mod Phys 88:025003. https://doi.org/10.1103/RevModPhys.88.025003
20. Levchenko A, Kamenev A (2008) Coulomb drag in quantum circuits. Phys Rev Lett 101:216806. https://doi.org/10.1103/PhysRevLett.101.216806
21. Raichev OE, Gusev GM, Hernandez FGG, Levin AD, Bakarov AK (2020) Phonon drag thermoelectric phenomena in mesoscopic two-dimensional conductors: current stripes, large Nernst effect, and influence of electron-electron interaction. Phys Rev B 102:195301. https://doi.org/10.1103/PhysRevB.102.195301
22. Strait JH, Holland G, Zhu W, Zhang C, Ilic BR, Agrawal A, Pacifici D, Lezec HJ (2019) Revisiting the photon-drag effect in metal films. Phys Rev Lett 123:053903. https://doi.org/10.1103/PhysRevLett.123.053903
23. Roussel B, Cabart C, Fève G, Thibierge E, Degiovanni P (2017) Electron quantum optics as quantum signal processing. Phys Status Solidi B 254:1600621. https://doi.org/10.1002/pssb.201600621
24. Litinski D, Brouwer PW, Filippone M (2017) Interacting mesoscopic capacitor out of equilibrium. Phys Rev B 96:085429. https://doi.org/10.1103/PhysRevB.96.085429
25. Idrisov EG, Levkivskyi IP, Sukhorukov EV (2018) Dephasing in a Mach-Zehnder interferometer by an Ohmic contact. Phys Rev Lett 121:026802. https://doi.org/10.1103/PhysRevLett.121.026802
26. Zhang G, Gornyi IV, Spånslätt C (2022) Delta-T noise for weak tunneling in one-dimensional systems: interactions versus quantum statistics. Phys Rev B 105:195423. https://doi.org/10.1103/PhysRevB.105.195423
27. Gurevich VL, Muradov MI (2015) Drag of ballistic electrons by an ion beam. J Exp Theor Phys 121:998–1006. https://doi.org/10.1134/S1063776115130026

28. Filliger R, Reimann P (2007) Brownian gyrator: a minimal heat engine on the nanoscale. Phys Rev Lett 99:230602. https://doi.org/10.1103/PhysRevLett.99.230602
29. Chiang K-H, Lee C-L, Lai P-Y, Chen Y-F (2017) Electrical autonomous Brownian gyrator. Phys Rev E 96:032123. https://doi.org/10.1103/PhysRevE.96.032123

Part III
Fluctuations, Non-equilibrium Effects, and Future Directions

Chapter 5
Mesoscopic Heat Multiplier and Fractionalizer

Abstract This chapter explores the emergence of correlated states in Hall chiral quantum systems, focusing on their impact on thermal and electrical transport. We analyze a setup where an Ohmic contact is connected to multiple edge states, with one forming a loop. Surprisingly, this looped edge state carries more heat than expected in equilibrium, an effect linked to charge fluctuations in the Ohmic contact. To probe this phenomenon, we consider two types of thermometry: capacitive coupling and quantum point contacts (QPCs). While a capacitive probe fails to detect excess heat in equilibrium, a QPC reveals a fundamental violation of the Wiedemann-Franz law, which means the ratio of thermal to electrical conductance deviates from its expected universal value. This violation depends only on the resistance of the circuit and persists in various temperature conditions. We further analyze how heating the Ohmic contact affects transport and confirm our predictions through numerical calculations. The findings connect to broader physics, including Tomonaga-Luttinger liquids and fractional quantum Hall systems, offering new insights into heat transport and fluctuation effects in chiral electronic systems.

5.1 Nature of Correlated States

In Chap. 3, we focused on using an OC as a toy model to investigate the influence of edge reconstruction or disorder on thermal transport in quantum Hall samples. We applied a combination of Langevin equations and scattering theory to a system resembling the Caldeira-Leggett model, effectively modeling dissipation in chiral systems. We demonstrated the quantization of heat for these systems.

In the preceding chapter, we explored the disruption of universal heat flux quantization in nonchiral systems, as discussed in Paper II. We reported a negative heat drag effect that stems from additional correlations mediated by the degree of freedom of collective charge. Local heat measurements can reveal a non-quantum amount of heat current carried by edge states in a system that is globally at equilibrium. This discovery led to the current chapter, in which we analyze the electrical and thermal properties of these correlated states using a similar device. We introduce a minimal model of an OC to demonstrate its ability to host correlated states, arising from strong

F. Stäbler, *Dissipation, Interactions, and Heat Transport in Quantum Hall Devices*, Springer Theses, https://doi.org/10.1007/978-3-031-90671-8_5

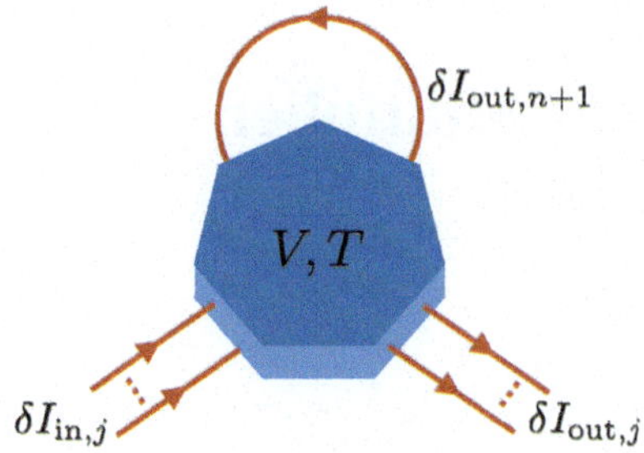

Fig. 5.1 An Ohmic contact connected to $n \geq 1$ external edge states and $m = 1$ looped edge state. This setup can be generalized to multiple loops. A setup like this was already experimentally realized in [1, 18]. The loop can be achieved by a quantum point contact (QPC) in the pinch off regime. As we will show in Sect. 5.4 at finite transmission the QPC can be used to probe the edge state. We consider the Ohmic contact at a uniform temperature T, but experimentally the contact might be heated via Joule heating. This situation can be treated under the assumption of local thermal equilibrium by taking different temperatures in the noise powers of the boundary currents and Langevin sources, see Sect. 5.4.3

Coulomb interactions, similar to the findings in [19, 20], as illustrated in Fig. 5.1. The OC is connected to macroscopically large electrodes through chiral edge states n, with an additional edge state arranged to form a loop that feeds back into the OC. The anomalous state appears in the looped edge state. The corresponding experimental setup closely resembles the one presented in [1, 18]. We show that these correlated states are related to voltage fluctuations of the OC, which is in full agreement with the P(E) theory [2], however the role of these extra fluctuations in electrical and thermal transport remained elusive in previous analyses. In the approach presented in this chapter we explicitly resolve the temperature dependence of the additional fluctuations, which had to be found with an experimental hypothesis in P(E) theory. We report a multiplication of heat, counterintuitively in the Coulomb blockade regime; A higher than quantum amount of heat carried by the interacting edge states, which poses a paradox of having a hot channel in a globally equilibrium system, similar to the discussions in Chap. 4. The paradox is resolved by coupling this interacting state to another state capacitively or via a quantum point contact (QPC) and explicitly showing energy conservation. We study the linear electrical conductance and thermal tunneling conductance and report a violation of the Wiedemann-Franz law by a number that only depends on the external resistance $R = \frac{R_q}{n}$ of the ohmic contact

$$\mathcal{L}_n = 3\frac{2+n}{2+3n}\mathcal{L}_0, \tag{5.1}$$

with the resistance quantum $R_q = 2\pi\hbar/e^2$. Furthermore, we study the Lorenz number in experimentally more realistic situations and numerically for arbitrary temperatures in equilibrium.

5.2 Multiplication of Heat Carried by an Edge State

Let us consider a generalization of the previously proposed model (2.9) consisting of N incoming and N outgoing edge channels to the Ohmic contact. The bosonic Hamiltonian of this system can be written as

$$\mathcal{H} = \frac{v_F}{4\pi}\sum_{i=1}^{2N}\int_{-\infty}^{\infty} dx\,(\partial_x\phi_i(x,t))^2 + \frac{Q^2(t)}{2C}, \tag{5.2}$$

where $i = 1, \ldots, N$ labels the incoming, and $i = N+1, \ldots, 2N$ the outgoing edge states and $Q(t) = \sum_{i'}\int_{-\infty}^{0} dx\,\rho_{i'}(x,t)e^{\frac{\varepsilon x}{v_F}}$ is the integral of the charge density $\rho_i(x,t) = \frac{e}{2\pi}\partial_x\phi_i(x,t)$ inside of the interaction region $x \in (-\infty, 0]$. The exponentially decaying term arises due to the finite lifetime of excitations ε inside of the Ohmic contact. A similar Hamiltonian was used in the references [21–23]. The Heisenberg equation of motion $\partial_t\phi_i(x,t) = -i\left[\phi_i(x,t), \mathcal{H}\right]$ can be recast into the form of Kirchoff equations (2.33a) and Langevin equations (2.33b) respectively[1]

$$\frac{d}{dt}\Delta Q(t) = \sum_{j=1}^{N}\left[\delta I_{\text{in},j}(t) - \delta I_{\text{out},j}(t)\right], \tag{5.3a}$$

$$\delta I_{\text{out},j}(t) = \frac{\Delta Q(t)}{\tau_C} + \delta I_{\text{oc},j}(t),. \tag{5.3b}$$

Next, consider taking m outgoing states and loop them back onto the ohmic contact, which reduces the number of unlooped external[2] states to $n = N - m$ incoming and outgoing states, as depicted in Fig. 5.1. This can be implemented by an additional constraint on the equation of motion Eqs. (5.3a) and (5.3b) of the form

$$\delta I_{\text{in},j'} = e^{i\frac{\omega L}{v_F}}\delta I_{\text{out},j'} \quad n < j' \leq N, \tag{5.4}$$

Our goal is to compute the heat flux carried by the looped edge states, which is expected to be anomalous, since additionally to the free fermionic neutral excitation carrying a heat flux quantum, we find the presence of some charge fluctuations adding to this heat flux.

To compute the heat flux carried by any edge state outgoing of the ohmic contact, we solve Eqs. (5.3a) and (5.3b) and express the outgoing currents in terms of incoming boundary currents and the Langevin sources. For simplicity, let us assume the system is in thermal equilibrium, i.e. the noise power of the incoming boundary current and the noise power of the Langevin sources are equal to an equilibrium noise power at

---

[1] Note that in contrast to Eq. (2.33b), we choose to display the "single channel" RC-time τ_C instead of the $\tau_{C,N} = \tau_C/N$. The latter depends on the number of channels attached to the OC, which we will vary.

[2] These states are referred to as external, since the series resistance of the circuit is given by $R = R_q/n$.

a known temperature $S_{\text{in}}(\omega) = S_{\text{oc}}(\omega) = S(\omega)$, given by Eq. (2.22). This constraint will be relaxed in Sect. 5.4.3. Furthermore, let us assume we look at the simple case of one loop $m = 1$ and n external channels and short loops $L = 0$, which corresponds to the experimental situation in [18]. For more general solutions, see Appendix C.1.

The noise power of the outgoing currents is defined as

$$S_{\text{out},j}(\omega) = 2\pi\delta(\omega + \omega')\left\langle \delta I_{\text{out},j}(\omega)\delta I_{\text{out},j}(\omega')\right\rangle, \tag{5.5}$$

and can be found by a Fourier transformation of Eqs. (5.3a), (5.3b) and (5.4). The noise power is given by

$$S_{\text{out},j}(\omega) = S(\omega), \quad 1 < j \leq n, \tag{5.6a}$$

$$S_{\text{out},n+1}(\omega) = \left(1 + \frac{2n}{n^2 + \omega^2\tau_C^2}\right) S(\omega), \tag{5.6b}$$

The expression for the noise power of the external channels is equilibrium as expected by earlier considerations, however the noise power of the loop acquires an additional contribution due to the charge fluctuations of the OC. According to Eq. (2.26) the heat flux carried by the looped edge state is given by

$$J_{\text{out},n+1} = J_q, \tag{5.7}$$

for high temperatures compared to the charging energy $\tau_C T \gg 1$ and

$$J_{\text{out},n+1} = \left(1 + \frac{2}{n}\right) J_q, \tag{5.8}$$

for low temperatures $\tau_C T \ll 1$. Let us stress again that the anomalous heat flux can be understood as a contribution from the fermionic neutral mode and charge fluctuations created by the Ohmic contact. This can be seen by decomposing the heat flux as

$$J_{\text{out},n+1} = \frac{R_q}{2}\left(\left\langle(\delta I_{\text{oc},n+1}(t))^2\right\rangle + \frac{1}{\tau_C^2}\left\langle(\Delta Q(t))^2\right\rangle\right), \tag{5.9}$$

where the first term yields a heat flux quantum, and the second term, by a simple counting argument in the strongly interacting limit, yields a factor of $2/n$ heat flux quanta. Solving the heat integral (2.26) numerically allows one to explore the intermediate values of J as a function of the channel number n and the dimensionless parameter $\tau_C T$. The graph in Fig. 5.2 shows the dependence of the heat flux for different values of the dimensionless inverse temperature and the number of channels.

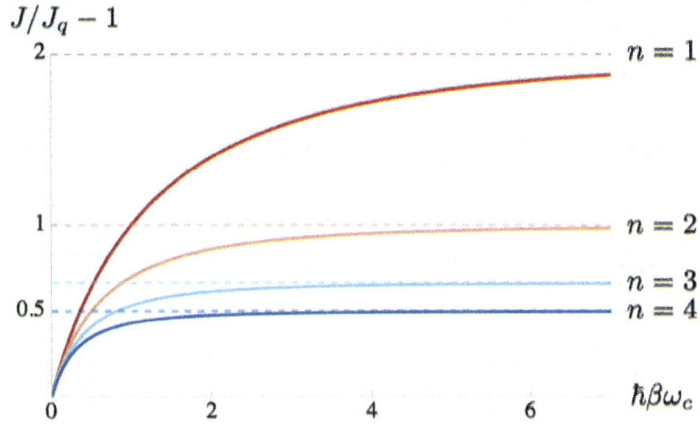

Fig. 5.2 The excess heat flux in the looped edge state normalized to a quantum $J/J_q - 1$ as a function of the dimensionless inverse temperature $\tau_C T$ for $n \in \{1, 2, 3, 4\}$ (solid lines) external edge channels. Note that increasing the number of edge channels decreases the magnitude of the excess heat flux as well as increasing the convergence towards the strongly interacting low temperature limit (dashed lines). Note that the plot presented here is taken from paper III and is plotted in units of $\hbar\beta\omega_c \equiv 1/(\tau_C T)$ in the notation of this paper

5.3 Edge State Thermometry Capacitive Coupling

5.3.1 *Equilibrium Operation*

In Sect. 5.2 we found that a mesoscopic device depicted by Fig. 5.1 shows a heat flux in the loop that is seemingly higher than the heat flux quantum. In other words, if the device is in global thermal equilibrium, that is, the temperature of the noise power of the incoming currents and sources is all equal to the base temperature T, the heat flux carried by the looped channel (5.8) could also be understood in terms of an effective temperature of the channel $T_{\text{eff}} \sim \sqrt{1 + \frac{2}{n}T}$ that is higher than the base temperature. Consider a thermometer probe attached to this looped state.

In a typical probe, a temperature difference would result in a net energy flux, as highlighted by [24]. However, in this case, the observed enhancement arises from charge fluctuations within the Ohmic reservoir. Nonetheless, it is crucial to address this paradox, given that extracting energy from a system in global equilibrium would violate the second law of thermodynamics. Adding the capacitively coupled channel as a thermometer can be implemented by adding the following equations to Eqs. (5.3a), (5.3b) and (5.4) (Fig. 5.3).

$$\begin{pmatrix} \delta I_{\text{out,th}} \\ \delta I_{\text{in},n+1} \end{pmatrix} = \begin{pmatrix} r_{\text{th}} & t_{\text{th}} \\ t_{\text{th}} & r_{\text{th}} \end{pmatrix} \begin{pmatrix} \delta I_{\text{in,th}} \\ \delta I_{\text{out},n+1} \end{pmatrix}. \tag{5.10}$$

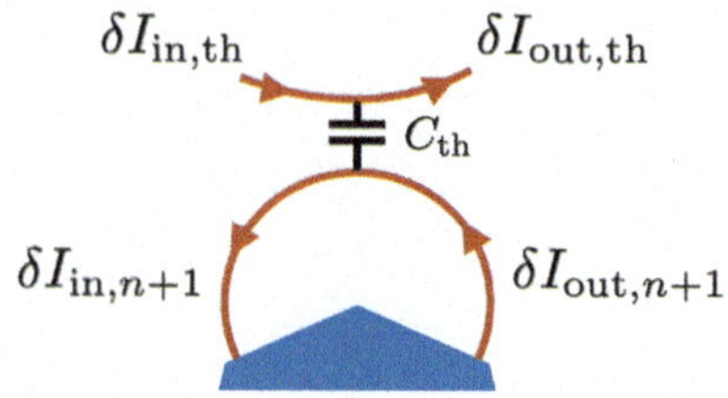

Fig. 5.3 Capacitive probe coupling the looped edge state to a free edge state. The thermometer can be operated in equilibrium or with a temperature difference between the edge states

The transmission and reflection amplitudes of the thermometer are given by

$$t_{\text{th}} = \frac{1 - 2e^{i\omega t_W} + e^{2i\omega t_W}}{2 - 2e^{i\omega t_W} - i\omega\tau_{\text{th}}}, \tag{5.11}$$

$$r_{\text{th}} = \frac{1 - e^{i\omega t_W} + ie^{i\omega t_W}\omega\tau_{\text{th}}}{2 - 2e^{i\omega t_W} + i\omega\tau_{\text{th}}}, \tag{5.12}$$

where $t_W = W/v_F$ is the time of flight through the probe and $\tau_{\text{th}} = R_q C_{\text{th}}$ is the RC-time of the capacitive interaction strength between the thermometer and the interacting mode, see Appendix C.2 for details.

In a next step, we compute the heat flux in the outgoing part of the thermometer to check if heat is carried over from the hot channel to the cold channel, using the same protocol described earlier, we evaluate the heat integral (2.22). Due to the a similar universality that was explored in earlier chapters, we find the expected result

$$\left\langle \delta I_{\text{out,th}}(\omega)\delta I_{\text{out,th}}(\omega')\right\rangle = 2\pi\delta(\omega+\omega')S_{\text{th}}(\omega) = S(\omega). \tag{5.13}$$

This clarifies the paradox: no heat extraction is possible in an equilibrium system due to the exact cancelation of additional correlations contributing to the heat flux in the outgoing heat flux calculation of the thermometer. This phenomenon arises because the capacitive thermometer is invasive by nature. Its capacitive coupling has dynamic and frequency-dependent characteristics, even at the lowest order of the coupling constant. The incoming state of the thermometer modifies the state of the Ohmic contact, which, in turn, alters the equation of motion for the outgoing thermometer state, effectively hindering heat extraction. Thus, we propose a truly noninvasive measurement technique using a tunnel junction, as detailed in Sect. 5.4.

5.3.2 *Non-equilibrium Operation*

Although correlations within the loop are not readily observable under equilibrium conditions, it is feasible to identify signatures of these states when the system is driven out of equilibrium. This can be achieved by introducing a temperature difference across the thermometer, typically through Joule heating of the OC. This results in different effective temperatures for the noise powers of incident current fluctuations

T, the Langevin sources at the contact $T_{\rm oc}$, and, for completeness, we also consider a separate temperature for fluctuations incoming from the thermometer's attached edge state $T_{\rm th}$. Consequently, we define noise power current fluctuations as having an equilibrium form but with a distinct temperature: $S_{\rm in,th} = S(T_{\rm th})$, $S_{\rm in} = S(T)$, and $S_{\rm oc} = S(T_{\rm oc})$, with

$$\langle \delta I_p(\omega)\delta I_{p'}(\omega')\rangle = 2\pi\delta_{pp'}\delta(\omega+\omega')S_p(\omega), \quad p \in \{\text{in}, \text{oc}, (\text{in}, \text{th})\} \tag{5.14}$$

A thermometer measurement should be non-invasive, leading us to consider a scenario where the interacting distance W is short, satisfying $t_W T_p \ll 1$ with $p \in \{\text{in}, \text{oc}, (\text{in}, \text{th})\}$. Under these conditions, we determine the noise power of the outgoing current fluctuations of the thermometer $S_{\rm th}(\omega)$, to be

$$\begin{aligned} S_{\rm th}(\omega) \approx S_{\rm in,th}(\omega) + \frac{\tau_C^2}{\tau_{\rm th}^2}\frac{\omega^2 t_W^4}{n^2+\omega^2\tau_C^2}\Big[\omega^2(S_{\rm oc}(\omega) - S_{\rm in,th}(\omega)) \\ + \frac{n}{\tau_C^2}((1+n)S_{\rm oc}(\omega) + S_{\rm in}(\omega) - (2+n)S_{\rm in,th}(\omega))\Big], \end{aligned} \tag{5.15}$$

By plugging this noise power into the heat integral (2.26) and subtracting the heat flux injected into the thermometer, we calculate the heat flux passing through the thermometer, $\delta J_{\rm th}$, which is measurable downstream of the thermometer edge state. In the limit where interactions are negligible, $\tau_C T_p \gg 1$, we find

$$\delta J_{\rm th} = \frac{\pi^3 t_W^4}{30\tau_{\rm th}^2}\left(T_{\rm oc}^4 - T_{\rm th}^4\right), \tag{5.16}$$

and in the strongly interacting limit $\tau_C T_p \ll 1$ we find

$$\delta J_{\rm th} = \frac{\pi^3 t_W^4}{30 n\tau_{\rm th}^2}\left(T^4 + (1+n)T_{\rm oc}^4 - (2+n)T_{\rm th}^4\right). \tag{5.17}$$

This highlights a dependency on the number of external edge states connected to the Ohmic contact. The presence of the correlated state within the loop has the potential to modify the (non-universal) prefactor of the heat flux transferred to the thermometer, compared to the non-interacting scenario, though it does not alter the overall temperature-dependent scaling. Additionally, it is important to mention that the outgoing thermometer noise power can be derived under different conditions or approached non-perturbatively. However, such analyses exceed the scope of this chapter and will be addressed in future work.

5.4 Breakdown of Wiedemann-Franz Law

In the last section we showed that in an equilibrium system it is impossible to see the special features of the looped channel. In non-equilibrium scenarios a capacitive coupling can reveal non-trivial features of these states; however, according to Eqs. (5.11) and (5.12) this coupling is a dynamical (frequency dependent) effect.

In this chapter, we explore a different type of thermometer, specifically one that consists of a tunneling junction, such as that formed by a Quantum Point Contact (QPC). Unlike the capacitive coupling, here we assume the transmission amplitudes to be small and frequency-independent. We anticipate the looped states to exhibit an anomalous fermionic correlation function, which significantly influences the electrical and thermal transport quantities discussed in this chapter. Using a tunneling Hamiltonian framework, we specifically address nonequilibrium scenarios by examining the linear response of both current and heat current to a small bias voltage δV and temperature difference δT.

5.4.1 Excursion on Non-equilibrium Bosonization

One of the most effective methods to address this problem is the non-equilibrium bosonization technique, as detailed in [25]. In this section, we provide a brief summary of the results to serve as both a toolbox and a starting point to understand the content of this chapter. Although we focus on clarity and utility, we do not claim that this summary is complete. For more comprehensive details, see the Appendix D.

At the heart of the bosonization procedure lies the non-linear mapping between fermionic and bosonic operators, known as the bosonization identity [26].

$$\psi(x,t) = \frac{1}{\sqrt{2\pi a}} e^{i\phi(x,t)}. \tag{5.18}$$

It can be shown that this identity fulfills the fermionic anti-commutation relation, guaranteeing the consistency of the approach. Additionally, this approach allows the representation of most Hamiltonians featuring density-density interactions, such as (2.9) by utilizing the point-splitting technique, as detailed in [27]. The strength of this method lies in its ability to transform certain interacting fermionic theories into a non-interacting bosonic theory.

Unlike "normal" bosonization, nonequilibrium bosonization leverages the fact that, in open systems, the bosonic fields can be expressed, via scattering theory, through boundary fields with well-characterized spectral properties. For example, these include spectral densities at different temperatures, thereby allowing the extension of bosonization techniques to address inherently non-equilibrium problems. To demonstrate the technique, we show the derivation of the fermionic correlation function through its bosonic counterpart.

$$\left\langle \psi^\dagger(t)\psi(t')\right\rangle = \frac{e^{i\mu(t-t')}}{2\pi a}\left\langle e^{-i\phi(t)}e^{i\phi(t')}\right\rangle, \tag{5.19}$$

where, we explicitly singled out the zero mode, that in the present case appears as an average bias μ of the edge state[3]. In a next step, we use the gaussianity of the theory and the fact that the bosonic correlation function has time translational invariance. This allows us to write

$$\left\langle e^{-i\phi(t)}e^{i\phi(t')}\right\rangle = e^{-\langle(\phi(0)-\phi(t-t'))\phi(0)\rangle} \tag{5.20}$$

Let us define the bosonic phase correlation function as $\ln C(t) = -\langle(\phi(0)-\phi(t))\phi(0)\rangle$. It is closely related to the noise power in the following way.

$$\ln C(t) = -2\pi\int \frac{d\omega}{\omega^2}S(\omega)(1-e^{-i\omega t}), \tag{5.21}$$

In this step the non-equilibrium bosonization enters, since the noise power $S(\omega)$ in the phase correlation function (5.21) depends on the boundary conditions. For the case of free bosons, where the noise power is given by its equilibrium form (2.22), we can evaluate the frequency integral with the following steps: (i) Express the noise power as a geometric series: (ii) Single out the zero temperature contribution and impose a high-frequency cutoff, given by the same UV cutoff we introduced in Eq. (5.18); (iii) Solve the remaining temperature dependent part and perform the sum. This gives

$$\begin{aligned} -\ln C(t) = 2\pi\int\frac{d\omega}{\omega}\frac{1-e^{-i\omega t}}{1-e^{-\omega/T}} &= \int_0^\infty \frac{d\omega}{\omega}(1-e^{-i\omega t})e^{-\frac{a\omega}{v_F}} \\ &+\sum_{n=1}^\infty\int_0^\infty\frac{d\omega}{\omega}(2-2\cos(\omega t))e^{-\omega n/T} \\ &= \log\left(\frac{iv_Ft+a}{a}\right)+\sum_{n=1}^\infty\log\left(1+\frac{T^2t^2}{n^2}\right) \\ &= \log\left(\left(\frac{iv_Ft+a}{a}\right)\prod_{n=1}^\infty\left(1+\frac{T^2t^2}{n^2}\right)\right) \\ &= \log\left(\left(\frac{iv_Ft+a}{a}\right)\frac{\sinh(\pi Tt)}{\pi Tt}\right). \end{aligned} \tag{5.22}$$

Putting everything together yields the free fermion correlation function

$$\left\langle \psi^\dagger(t)\psi(t')\right\rangle = -\frac{iT}{2v_F}\frac{e^{i\mu(t-t')}}{\sinh(\pi T(t-t'-i\eta))}, \tag{5.23}$$

[3] Note that $\left\langle\psi(t)\psi^\dagger(t')\right\rangle = e^{-i\mu(t-t')}\left\langle e^{i\phi(t)}e^{-i\phi(t')}\right\rangle$ appears with the opposite sign.

where the shift of the pole is chosen such that we obtain the correct Fermi distribution function if the expression is Fourier transformed and especially in the zero temperature limit $T \to 0$ we find

$$\left\langle \psi^\dagger(t)\psi(t')\right\rangle = -\frac{i}{2v_F}\frac{e^{i\mu(t-t')}}{t-t'-i\eta}. \tag{5.24}$$

The correlation function of the conjugated therm can be found according to Eqs. (5.28) and (5.29).

5.4.2 *Lorenz Number of Tunneling Current*

Let us assume a coupling of a looped edge state to a free edge state via a quantum point contact (QPC), see Fig. 5.4. In the following, we use the tunneling Hamiltonian approach. Consider two counterpropagating chiral edge states interconnected by tunneling at position x_0. The Hamiltonian of the system can be represented as follows.

$$\mathcal{H} = \sum_{\sigma=\pm} \mathcal{H}_\sigma + \mathcal{H}_T \tag{5.25a}$$

$$\mathcal{H}_\sigma = -i\sigma v_F \int_{-\infty}^{\infty} dx\, \psi_\sigma^\dagger(x,t)\partial_x\psi_\sigma(x,t), \tag{5.25b}$$

$$\mathcal{H}_T = \gamma\psi_+^\dagger(x_0,t)\psi_-(x_0,t) + \text{h.c.}. \tag{5.25c}$$

This formalism treats tunneling separately from the interations in th Ohmic contact, allowing to have fermionic boundary conditions for the free edge state ψ_+ and more complex boundary conditions for ψ_- due to interactions at the Ohmic contact. For details of the calculations, please see Appendix C.3. The average current and heat flux are determined by the rate of change in charge and energy in the free system, respectively. This leads to the following expression for the average tunneling current:

$$\begin{aligned} I_{\text{tun}} =& \gamma^2 \int_{-\infty}^{\infty} dt \left(\left\langle \psi_+(t)\psi_+^\dagger(0)\right\rangle\left\langle \psi_-^\dagger(t)\psi_-(0)\right\rangle\right. \\ &\left. - \left\langle \psi_+^\dagger(0)\psi_+(t)\right\rangle\left\langle \psi_-(0)\psi_-^\dagger(t)\right\rangle\right), \end{aligned} \tag{5.26}$$

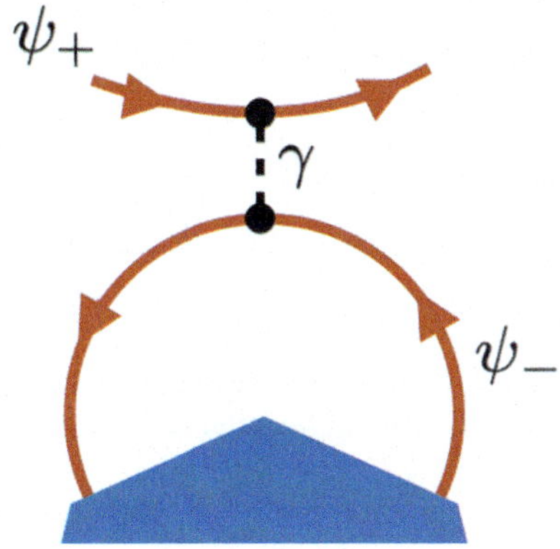

Fig. 5.4 In our analysis of the Quantum Point Contact (QPC), we use the tunneling Hamiltonian formalism. This approach models the process where an electron is created in the "+" edge and annihilated in the "−" edge, or vice versa, mediated by a perturbatively small interaction constant, γ. The tunneling interaction is switched on adiabatically, allowing us to consider it independently from the Coulomb interactions present in the Ohmic contact. We then proceed to calculate the electrical and thermal conductance at leading order in the tunneling constant

and for the tunneling heat flux

$$J_{\text{tun}} = -i\gamma^2 \int_{-\infty}^{0} dt \Big(\big\langle \psi_-^\dagger(0)\psi_-(t)\big\rangle \big\langle \dot{\psi}_+(0)\psi_+^\dagger(t)\big\rangle - \big\langle \psi_-(t)\psi_-^\dagger(0)\big\rangle \big\langle \psi_+^\dagger(t)\dot{\psi}_+(0)\big\rangle \Big), \tag{5.27}$$

where $\dot{\psi}(t') = \lim_{t\to t'} \partial_t \psi(t)$.

The free fermionic correlation functions (of ψ_+) were calculated in the previous section and are given by Eq. (5.23). The interacting correlation functions (of ψ_-) can be calculated in the same way by replacing $S(\omega)$ in Eq. (5.21) by the modified noise power $S_{\text{out},n+1}(\omega)$ given by Eq. 5.6b. In other words, every two-point function can be expressed as

$$\big\langle \psi_\sigma(t)\psi_\sigma^\dagger(t')\big\rangle = e^{-i\mu_\sigma(t-t')}C_\sigma(t-t'), \tag{5.28}$$

$$\big\langle \psi_\sigma^\dagger(t)\psi_\sigma(t')\big\rangle = e^{i\mu_\sigma(t-t')}C_\sigma(t-t'), \tag{5.29}$$

where the first exponential function is the zero-mode contribution and the phase correlation function $\ln C_\sigma(t) = \langle(\phi_\sigma(0) - \phi_\sigma(t))\phi_\sigma(0)\rangle$ is related to the noise power via

$$\ln C_\sigma(t) = -2\pi \int \frac{d\omega}{\omega^2} S_\sigma(\omega)(1 - e^{-i\omega t}), \tag{5.30}$$

where $S_+(\omega) = S(\omega)$ and $S_-(\omega) = S_{\text{out},n+1}(\omega)$. The free fermionic phase correlation function is given by

$$C_+(t) = -\frac{iT}{2v_F} \frac{1}{\sinh\left(\pi T \left(t - i\eta\right)\right)}, \tag{5.31}$$

where the shift of the pole was introduced to find the correct Fourier transformation in terms of the occupation number. For the interacting phase correlation function we find it by relying on a separation of energy scales. The correlation function is altered by the interaction parameter τ_C. For temperatures much larger than the charging energy $\tau_C T \gg 1$, we know that our correlation function must have free fermionic character, e.g. due to Eq. 5.7. For temperatures much smaller than the charging energy $\tau_C T \ll 1$, which is in turn much smaller than the UV cutoff, we can find an asymptotic solution in the long time limit $t/\tau_C \gg 1$ of the form

$$C_-(t) = \frac{\left(\frac{\tau_C \pi e^{-\gamma_{\mathrm{EM}}}}{n}\right)^{\frac{2}{n}}}{2v_F} \left(\frac{-iT}{\sinh\left(\pi T \left(t - t' - i\eta\right)\right)}\right)^{1+\frac{2}{n}} \tag{5.32}$$

see Appendix C.4 for a detailed derivation.

This allows us to explicitly compute the tunneling current (5.26) and the tunneling heat flux (5.27) explicitly. Let us start from the average current. Details of the calculations can be found in Appendix C.5; we will give the results assuming a separation of energy scales given by $T, t^{-1} \ll \tau_C^{-1} \ll v_F a^{-1}$.

5.4.2.1 Linear Conductance

The linear conductance, assuming a small voltage bias across the QPC, is given by

$$G_n = \frac{\gamma^2}{2\pi v_F^2} \left(\frac{\tau_C T \pi e^{-\gamma_{\mathrm{EM}}}}{n}\right)^{\frac{2}{n}} \frac{\sqrt{\pi}\,\Gamma\left(1 + \frac{1}{n}\right)}{2\Gamma\left(\frac{3}{2} + \frac{1}{n}\right)}, \tag{5.33}$$

which in general depends on the number of channels n and reproduces the free fermionic result in the limit of $n \to \infty$

$$G_\infty = \frac{\gamma^2}{2\pi v_F^2}. \tag{5.34}$$

5.4.2.2 Heat Flux

It becomes apparent that the heat flux is zero when the temperatures of the two terminals are equal. For a small temperature difference δT across the QPC, the linear response heat flux is given by

$$J_n = \frac{4^{\frac{1}{n}} \left((n-1)\Gamma\left(2+\frac{1}{n}\right)^2 - \Gamma\left(1+\frac{1}{n}\right)\Gamma\left(3+\frac{1}{n}\right)\right)}{n\Gamma\left(4+\frac{2}{n}\right)} \times \left(\frac{\tau_C T \pi e^{-\gamma_{EM}}}{n}\right)^{\frac{2}{n}} \frac{\pi T \delta T \gamma^2}{v_F^2}, \tag{5.35}$$

Especially for $n \to \infty$, we find the expected free fermionic heat flux

$$J_\infty = \kappa_0 T \delta T \frac{\gamma^2}{v_F^2}, \tag{5.36}$$

with $\kappa_0 = \pi/6$ being the thermal conductance quantum. The Lorenz number for a number of channels n is obtained through some remarkable cancellations as

$$\mathcal{L}_n = \frac{J_n}{T\delta T G_n} = \frac{2+n}{2+3n}\pi^2 = 3\frac{2+n}{2+3n}\mathcal{L}_\infty, \tag{5.37}$$

where we defined the non-intercting Lorenz number $\mathcal{L}_\infty = \pi^2/3$. Note that the same result may be obtained by considering transport integrals [28, 29].

5.4.3 *Lorenz Number for Selective Temperature Bias*

In actual experiments, such as those studying dynamical Coulomb blockade under a temperature bias [1, 2], the Ohmic contact is subjected to Joule heating. This can result in a nonuniform temperature distribution across the device, where the base temperature and the temperatures of the reservoir sources may differ. Under these conditions, the noise power in the loop is not determined by Eq. (5.6b), but rather by

$$S_{\text{out},n+1}(\omega) = S_{\text{oc}}(\omega) + \frac{n\left(S_{\text{oc}}(\omega) + S_{\text{in}}(\omega)\right)}{n^2 + \omega^2\tau_C^2}, \tag{5.38}$$

where the noise power is given by Eq. (2.22) with the respective temperatures T_{oc} of the Langevin sources and T_{in} for the current fluctuations incident to the device. Note that the noise power (5.38) and the structure of the correlation function in the loop (5.9) implies that our theory can be viewed and is equivalent to the commonly used $P(E)$ theory. Tunneling to the loop is thus equivalent to tunneling directly to the Ohmic contact.

The complex temperature dependence of the noise power directly influences the phase correlation function and consequently the Lorenz number. Consider a scenario where only the Ohmic contact is heated, while the other channels remain at the base temperature, that is, $T_{\text{in}} = T_+ = T$, where T_+ denotes the temperature of the Fermi distribution for the ψ_+ fermions. Under these conditions, the correlation function, given by Eq. 5.32, is

$$C_-(t) = \frac{\left(\frac{\tau_C \pi e^{-\gamma_{\mathrm{EM}}}}{n}\right)^{\frac{2}{n}}}{2v_F} \left(\frac{-iT}{\sinh\left(\pi T\left(t-t'-i\eta\right)\right)}\right)^{\frac{1}{n}} \times \left(\frac{-iT_{\mathrm{oc}}}{\sinh\left(\pi T_{\mathrm{oc}}\left(t-t'-i\eta\right)\right)}\right)^{1+\frac{1}{n}}, \tag{5.39}$$

where we assume $T_{\mathrm{oc}} = T + \delta T$ in the linear response regime and repeat the calculations that lead to the heat flux Eq. (5.35). This leads to a modified Lorentz number given by

$$\mathcal{L}_n^{\mathrm{sel.}} = \left(1 + \frac{1}{2+3n}\right)\mathcal{L}_\infty. \tag{5.40}$$

The maximum Lorenz number is reduced from $\mathcal{L}_1 = \frac{9}{5}$ to $\mathcal{L}_1^{\mathrm{sel.}} = \frac{6}{5}$.

In [2], the authors noted that after Joule heating, the temperature of the node differs from that of the electromagnetic environment. The closest match to the experimental data was achieved using the mean temperature, $(T_{\mathrm{in}} + T_{\mathrm{oc}})/2$, with small deviations. We can demonstrate, using (5.38), that this is true for leading order for small temperature differences between the node and its environment. This is a strong advantage of the nonequilibrium bosonization approach [25], which allows a more intuitive physical picture to be produced, by describing the voltage fluctuations in terms of boundary currents. This allows us to derive the aforementioned temperature of the voltage fluctuations rigorously. We can estimate the deviation from the environmental temperature $T_{\mathrm{env}} = (T_{\mathrm{in}} + T_{\mathrm{oc}})/2$, by comparing it with the exact solution of the tunneling current (5.26) using the exact correlation function (5.39). We can compute the tunneling conductance under an arbitrary temperature bias (assuming $T_+ = T_{\mathrm{in}}$) as

$$\frac{G_n}{G_\infty} = \left(\frac{\pi^2 e^{-\gamma_{\mathrm{EM}}}}{2E_C n}\right)^{\frac{2}{n}} \int_{-\infty}^{\infty} dt \sum_{\sigma=\pm1} \frac{\pi t}{2i\sigma} \times \left(\frac{\sigma i T_{\mathrm{in}}}{\sinh(\pi T_{\mathrm{in}}(t+\sigma i\eta))}\right)^{1+\frac{1}{n}} \left(\frac{\sigma i T_{\mathrm{oc}}}{\sinh(\pi T_{\mathrm{oc}}(t+\sigma i\eta))}\right)^{1+\frac{1}{n}}. \tag{5.41}$$

In contrast to that, the approximation used in [2] is given by

$$\frac{G_{\mathrm{env}}}{G_\infty} = \left(\frac{\pi^2 e^{-\gamma_{\mathrm{EM}}}}{2E_C n}\right)^{\frac{2}{n}} \int_{-\infty}^{\infty} dt \sum_{\sigma=\pm1} \frac{\pi t}{2i\sigma} \left(\frac{\sigma i T_{\mathrm{in}}}{\sinh(\pi T_{\mathrm{in}}(t+\sigma i\eta))}\right) \times \left(\frac{\sigma i T_{\mathrm{oc}}}{\sinh(\pi T_{\mathrm{oc}}(t+\sigma i\eta))}\right) \left(\frac{\sigma i T_{\mathrm{env}}}{\sinh(\pi T_{\mathrm{env}}(t+\sigma i\eta))}\right)^{\frac{2}{n}}. \tag{5.42}$$

A direct comparison between the exact conductance formula and the approximation made in the earlier works can be seen in Fig. 5.5, which is remarkably similiar to Fig. 8 in [2] and confirms the experimental hypothesis that $T_{\mathrm{env}} \gtrsim (T_{\mathrm{in}} + T_{\mathrm{oc}})/2$. The

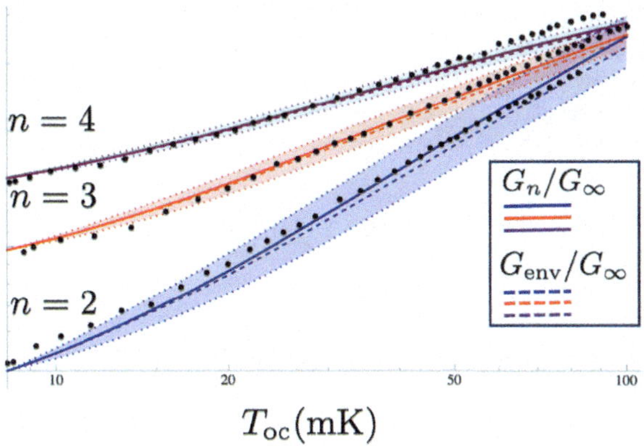

Fig. 5.5 Plot of the relative conductance (5.42) with the dashed line being $T_{\text{env}} = (T_{\text{in}} + T_{\text{oc}})/2$, the shaded regions are $T_{\text{env}} \in (T_{\text{in}}, T_{\text{oc}})$ and the full lines are the exact temperature dependence based on the conductance (5.41). The black dots are the experimental data from [2], provided by the experimentalists. The experimental error seems to be larger than the deviation of the exact from the approximate solution, which results in only a qualitative improvement

deviations of the exact solution from the approximate one are of the order of 1-5%, which seems to be consistent with the measurement within experimental errors.

5.4.4 Temperature Dependence of the Lorenz Number

If the separation of energy scales $T, t^{-1} \ll \tau_C^{-1} \ll v_F a^{-1}$ is not fulfilled, it is possible to study the exact solution for linear conductance and thermal conductance numerically.[4] In the following, we plot the Lorenz number for a device with $T_{\text{in}} = T_{\text{oc}}$ using numerical solutions derived from Eqs. (5.26) and (5.27) and the exact correlation function, as also referenced in [2].

$$\frac{\mathcal{L}_n}{\mathcal{L}_\infty} = \frac{\int_0^\infty dx \; \frac{3}{2}(3 - \cosh(2x))\text{sech}^4(x)e^{\Phi(x)}}{\int_0^\infty dx \; \text{sech}^2(x)e^{\Phi(x)}}, \tag{5.43}$$

with the hyperbolic secant function $\text{sech}(x) = (\cosh(x))^{-1}$.

[4] We assume that the tunneling strength γ represents the smallest energy scale; otherwise, other perturbations may become relevant. For further details on this consideration, see the discussion in [33].

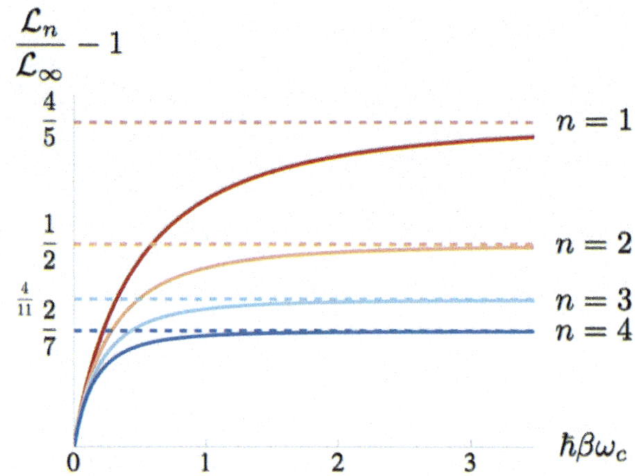

Fig. 5.6 Numerical values of the Lorenz number for $n \in \{1, 2, 3, 4\}$ are shown with solid lines. The dashed lines indicate the theoretical predictions for the low-temperature, strongly interacting regime. Note that the plot presented here is taken from paper III and is plotted in units of $\hbar\beta\omega_c \equiv 1/(\tau_C T)$ in the notation of this paper

$$\Phi(x) = -\frac{2x}{n} + \frac{\pi\left(\cos\left(\frac{n}{2\tau_C T}\right) - e^{-\frac{nx}{\pi\tau_C T}}\right)}{n\sin\left(\frac{n}{2\tau_C T}\right)} - \frac{2\log\left(1+e^{-2x}\right)}{n}$$
$$-\frac{1}{n}\sum_{\sigma=\pm1} H_{\frac{\sigma n}{2\pi\tau_C T}} + e^{-i\frac{\sigma n}{2\tau_C T}+\frac{\sigma n x}{\pi\tau_C T}} B\left(-e^{-2x}; 1 + \frac{\sigma n}{2\pi\tau_C T}, 0\right), \tag{5.44}$$

where H_α is a harmonic number and $B(x; a, b)$ is the incomplete Beta function. The ratio of the Lorenz number to one is plotted in Fig. 5.6.

Note that a large number of channels suppresses the magnitude of the Lorenz number, i.e. the effect of voltage fluctuations, but also leads to faster convergence towards the strongly interacting prediction, as indicated by the dashed lines, upon lowering the temperature. The Lorenz number for arbitrary temperatures can be studied in a more cumbersome but similar manner by deriving the equivalent expression of the numerical integral (5.43) using the general expression for the noise power (5.38).

5.4.5 Weak Backscattering Limit

If the QPC is operated in the weak tunneling regime, as depicted in Fig. 5.7, we predict a Lorenz number that is determined by mapping $n \to -(n+1)$, according to Appendix C.3.3. This is consistent with the duality in the Luttinger parameter discussed in [2]. The Lorenz number in the weak backscattering regime can be expressed as follows:

$$\mathcal{L}_n = \frac{3(1-n)}{2-3(1+n)}\mathcal{L}_\infty. \tag{5.45}$$

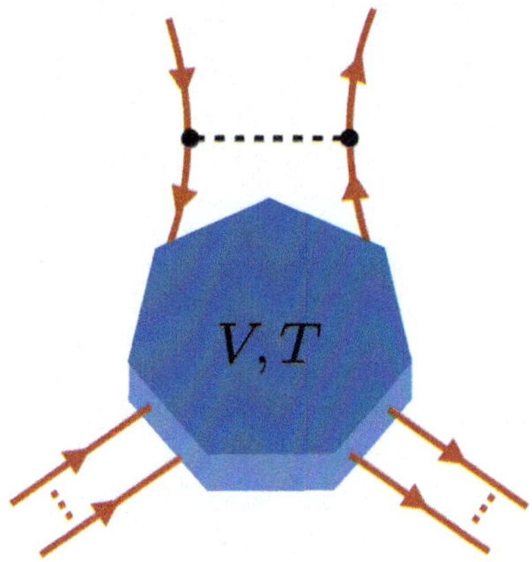

Fig. 5.7 In the weak backscattering configuration of the QPC, the Lorenz number is expected to be dual to the weak tunneling case

5.4.6 Additional Comments

The Lorenz number presented in this paper was calculated for an experimentally established set-up, where the external resistance can be altered by changing the series resistance of the Ohmic contact in discrete steps, represented as $R = R_q/n$. However, it is also possible to consider a circuit in series with a macroscopic resistor whose resistance can be adjusted freely. The analyses in this paper do not depend on n being an integer, allowing us to extend our results to cover any arbitrary series resistance R for the Lorenz number.

$$\mathcal{L}_n = 3\frac{2 + R_q/R}{2 + 3R_q/R}\mathcal{L}_\infty. \tag{5.46}$$

As demonstrated in [2, 30–32], there exists a direct mapping between an electronic channel in series with a linear resistance and a Tomonaga-Luttinger liquid (TLL) model for an infinitely long 1D system of spinless electrons interacting with a single impurity. The interaction parameter of the TLL is given by

$$K = \left(1 + \frac{R}{R_q}\right)^{-1}, \tag{5.47}$$

which extends our results directly to the TLL case mentioned above.

A violation of the Wiedemann-Franz law and the expected thermopower was also reported in interacting Luttinger liquids by [33, 34], but the reported dependence of the prefactor on the interaction parameter is different.

Lastly, note that the Lorenz number, given by Eq. (5.37) for $n \to 1$, also appears in the context of electron tunneling of fractional quantum Hall (FQH) quasi-particles $\nu = 1/3$, as discussed in [35], however, is usually strongly affected by interactions [36, 37]. Our device is similar to the fractionalizer mentioned in [20], which generates anyonic excitations. Many interpretations regarding the nature of these states are applicable to our study as well.

5.5 Conclusion

In this chapter, insights are provided into the previously overlooked heat-multiplication effect within OCs. The enhancement of the locally measured heat flux manifests itself in the Coulomb blockade regime, which initially may seem counterintuitive. However, this is attributed to an additional contribution from the collective charge mode via the fluctuating potential $\Delta Q(t)/C$ of the OC. A capacitive probe and a quantum point contact were used for thermometry on this seemingly hotter edge state, thereby ruling out any paradoxes related to the violation of energy conservation.

One consequence of this investigation is the violation of the Wiedemann-Franz tunneling law, which leads to a new set of universal Lorenz numbers. These numbers, representing the relationship between linear thermal response and voltage response, depend solely on the series resistance R of the circuit. Initially, a separation of energy scales is assumed and the Lorenz number is studied in the Coulomb blockade regime, where the charging energy significantly exceeds the energy scale set by temperature. We analyze a uniformly heated and a selectively heated Ohmic contact with a combination of the Langevin equation and scattering theory. This approach resulted in a larger than expected Lorenz ratio for both scenarios. Our methodology offers a more intuitive understanding of the role of fluctuations in contrast to P(E) theory, particularly in the context of non-local heat transport as discussed in Chap. 4. Specifically, it facilitates the rigorous derivation of the experimental hypothesis regarding environmental temperature made in [2].

Numerical studies were performed to explore the temperature dependence of the Lorenz number beyond the Coulomb blockade regime. Furthermore, we showed how our model can be mapped to continuous resistances, a Tomonaga-Luttinger liquid model, and its applicability for fractional fillings. This methodology could serve as a tool to probe the edge structure of complex filling factors. Furthermore, we became aware of the work reported in [29], which also documents a similar violation of the Wiedemann-Franz law.

References

1. Sivre E, Duprez H, Anthore A, Aassime A, Parmentier FD, Cavanna A, Ouerghi A, Gennser U, Pierre F (2019) Electronic heat flow and thermal shot noise in quantum circuits. Nat Commun 10:5638. https://doi.org/10.1038/s41467-019-13566-8
2. Duprez H, Pierre F, Sivre E, Aassime A, Parmentier FD, Cavanna A, Ouerghi A, Gennser U, Safi I, Mora C, Anthore A (2021) Dynamical Coulomb blockade under a temperature bias. Phys Rev Res 3:023122. https://doi.org/10.1103/PhysRevResearch.3.023122
3. Narozhny BN, Levchenko A (2016) Coulomb drag. Rev Mod Phys 88:025003. https://doi.org/10.1103/RevModPhys.88.025003
4. Levchenko A, Kamenev A (2008) Coulomb drag in quantum circuits. Phys Rev Lett 101:216806. https://doi.org/10.1103/PhysRevLett.101.216806
5. Raichev OE, Gusev GM, Hernandez FGG, Levin AD, Bakarov AK (2020) Phonon drag thermoelectric phenomena in mesoscopic two-dimensional conductors: current stripes, large Nernst effect, and influence of electron-electron interaction. Phys Rev B 102:195301. https://doi.org/10.1103/PhysRevB.102.195301
6. Strait JH, Holland G, Zhu W, Zhang C, Ilic BR, Agrawal A, Pacifici D, Lezec HJ (2019) Revisiting the photon-drag effect in metal films. Phys Rev Lett 123:053903. https://doi.org/10.1103/PhysRevLett.123.053903
7. Roussel B, Cabart C, Fève G, Thibierge E, Degiovanni P (2017) Electron quantum optics as quantum signal processing. Phys Status Solidi 254:1600621. https://doi.org/10.1002/pssb.201600621
8. Litinski D, Brouwer PW, Filippone M (2017) Interacting mesoscopic capacitor out of equilibrium. Phys Rev B 96:085429. https://doi.org/10.1103/PhysRevB.96.085429
9. Gurevich VL, Muradov MI (2015) Drag of ballistic electrons by an ion beam. J Exp Theor Phys 121:998–1006. https://doi.org/10.1134/S1063776115130026
10. Ben-Abdallah P, Biehs SA, Joulain K (2011) Many-body radiative heat transfer theory. Phys Rev Lett 107:114301. https://doi.org/10.1103/PhysRevLett.107.114301
11. Bhandari B, Chiriacò G, Erdman PA, Fazio R, Taddei F (2018) Thermal drag in electronic conductors. Phys Rev B 98:035415. https://doi.org/10.1103/PhysRevB.98.035415
12. Ben-Abdallah P (2019) Thermal photon drag in many-body systems. Phys Rev B 99:201406. https://doi.org/10.1103/PhysRevB.99.201406
13. Berdanier W, Scaffidi T, Moore JE (2019) Energy drag in particle-hole symmetric systems as a quantum quench. Phys Rev Lett 123:246603. https://doi.org/10.1103/PhysRevLett.123.246603
14. Chiang K-H, Lee C-L, Lai P-Y, Chen Y-F (2017) Electrical autonomous Brownian gyrator. Phys Rev E 96:032123. https://doi.org/10.1103/PhysRevE.96.032123
15. Idrisov EG, Levkivskyi IP, Sukhorukov EV (2018) Dephasing in a Mach-Zehnder interferometer by an ohmic contact. Phys Rev Lett 121:026802. https://doi.org/10.1103/PhysRevLett.121.026802
16. Zhang G, Gornyi IV, Spånslätt C (2022) Delta-T noise for weak tunneling in one-dimensional systems: interactions versus quantum statistics. Phys Rev B 105:195423. https://doi.org/10.1103/PhysRevB.105.195423
17. Filliger R, Reimann P (2007) Brownian Gyrator: a minimal heat engine on the nanoscale. Phys Rev Lett 99:230602. https://doi.org/10.1103/PhysRevLett.99.230602
18. Sivre E, Anthore A, Parmentier FD, Cavanna A, Gennser U, Ouerghi A, Jin Y, Pierre F (2018) Heat Coulomb blockade of one ballistic channel. Nat Phys 14:145–148. https://www.nature.com/articles/nphys4280
19. Idrisov EG, Levkivskyi IP, Sukhorukov EV (2020) Quantum ammeter: measuring full counting statistics of electron currents at quantum timescales. Phys Rev B 101:245426. https://doi.org/10.1103/PhysRevB.101.245426
20. Morel T, Lee J-YM, Sim H-S, Mora C (2022) Fractionalization and anyonic statistics in the integer quantum Hall collider. Phys Rev B 105:075433. https://doi.org/10.1103/PhysRevB.105.075433

21. Matveev KA (1995) Coulomb blockade at almost perfect transmission. Phys Rev B 51:1743–1751. https://doi.org/10.1103/PhysRevB.51.1743
22. Furusaki A, Matveev KA (1995) Theory of strong inelastic cotunneling. Phys Rev B 52:16676–16695. https://doi.org/10.1103/PhysRevB.52.16676
23. Slobodeniuk AO, Levkivskyi IP, Sukhorukov EV (2013) Equilibration of quantum Hall edge states by an Ohmic contact. Phys Rev B 88:165307. https://doi.org/10.1103/PhysRevB.88.165307
24. Sukhorukov EV, Cheianov VV (2007) Resonant dephasing in the electronic Mach-Zehnder interferometer. Phys Rev Lett 99:156801. https://doi.org/10.1103/PhysRevLett.99.156801
25. Sukhorukov EV (2016) Scattering theory approach to bosonization of non-equilibrium mesoscopic systems. Phys E: Low-Dimens Syst Nanostructures 77:191–198. http://www.sciencedirect.com/science/article/pii/S1386947715302836
26. Giamarchi T (2004) Quantum physics in one dimension. Oxford University Press, Oxford, New York. https://academic.oup.com/book/27811
27. von Delft J, Schoeller H (1998) Bosonization for beginners—refermionization for experts. Annalen der Physik 510:225–305. https://onlinelibrary.wiley.com/doi/abs/10.1002/andp.19985100401
28. Karki DB (2020) Coulomb blockade oscillations of heat conductance in the charge Kondo regime. Phys Rev B 102:245430. https://doi.org/10.1103/PhysRevB.102.245430
29. Kiselev MN (2023) Generalized Wiedemann-Franz law in a two-site charge Kondo circuit: Lorenz ratio as a manifestation of the orthogonality catastrophe. Phys Rev B 108:L081108. https://doi.org/10.1103/PhysRevB.108.L081108
30. Anthore A, Iftikhar Z, Boulat E, Parmentier FD, Cavanna A, Ouerghi A, Gennser U, Pierre F (2018) Circuit quantum simulation of a Tomonaga-Luttinger liquid with an impurity. Phys Rev X 8:031075. https://doi.org/10.1103/PhysRevX.8.031075
31. Jezouin S, Albert M, Parmentier FD, Anthore A, Gennser U, Cavanna A, Safi I, Pierre F (2013) Tomonaga-Luttinger physics in electronic quantum circuits. Nat Commun 4:1802. https://www.nature.com/articles/ncomms2810
32. Safi I, Saleur H (2004) One-channel conductor in an ohmic environment: mapping to a Tomonaga-Luttinger liquid and full counting statistics. Phys Rev Lett 93:126602. https://doi.org/10.1103/PhysRevLett.93.126602
33. Kane CL, Fisher MPA (1996) Thermal transport in a Luttinger liquid. Phys Rev Lett 76:3192–3195. https://doi.org/10.1103/PhysRevLett.76.3192
34. Schulz HJ (1993) Interacting fermions in one dimension: from weak to strong correlation. arXiv: cond-mat/9302006
35. Nosiglia C, Park J, Rosenow B, Gefen Y (2018) Incoherent transport on the $\nu = 2/3$ quantum Hall edge. Phys Rev B 98:115408. https://doi.org/10.1103/PhysRevB.98.115408
36. Asasi H, Mulligan M (2020) Partial equilibration of anti-Pfaffian edge modes at $\nu = 5/2$. Phys Rev B 102:205104. https://doi.org/10.1103/PhysRevB.102.205104
37. Srivastav SK, Kumar R, Spånslätt C, Watanabe K, Taniguchi T, Mirlin AD, Gefen Y, Das A (2021) Vanishing thermal equilibration for Hole-Conjugate fractional quantum Hall states in graphene. Phys Rev Lett 126:216803. https://doi.org/10.1103/PhysRevLett.126.216803

Chapter 6
Temperature Fluctuations of QHE Channels in the Heat Coulomb Blockade Regime

Abstract This chapter explores heat noise power and its fluctuations in chiral quantum Hall systems, focusing on how finite capacitance and temperature fluctuations of the Ohmic contact (OC) influence thermal transport. We first show that in equilibrium the heat noise follows a universal form, while in the Coulomb blockade regime the OC suppresses heat equilibration. Introducing temperature fluctuations due to finite heat capacity leads to a new energy relaxation time, affecting both the heat noise and thermal conductance. We extend the analysis to multiple edge channels, demonstrating how the OC's finite capacitance alters heat dynamics. Temperature fluctuations also generate non-Gaussian features in charge transport, leading to a non-zero fourth cumulant of the transmitted charge, which is absent in purely quadratic bosonized models. Additionally, applying a voltage bias results in a non-zero third cumulant, further highlighting the role of OC temperature fluctuations in generating non-Gaussian effects. These findings provide a deeper understanding of heat noise in mesoscopic conductors, revealing how charge-heat coupling and slow temperature dynamics influence fluctuations beyond standard Gaussian approximations.

6.1 Heat Noise Power

This chapter extends beyond the discussions in Sect. 2 by focusing on the fluctuations of the heat current in the output channel. Our analysis begins with an observation regarding the charge-heat relation expressed in Eq. (2.16), which establishes a critical link between the equilibrium charge and heat noise powers, as detailed in [1]:

$$S^Q(\omega) = \frac{R_q^2}{4\pi} \int d\omega_1 S(\omega_1) S(\omega - \omega_1). \tag{6.1}$$

Upon performing the integration, as referenced in [2], we obtain:

$$S^Q(\omega) = \frac{\omega}{48\pi} \left[(2\pi T)^2 + \omega^2\right] \left(1 + \coth\left(\frac{\omega}{2T}\right)\right). \tag{6.2}$$

F. Stäbler, *Dissipation, Interactions, and Heat Transport in Quantum Hall Devices*, Springer Theses, https://doi.org/10.1007/978-3-031-90671-8_6

In particular, the heat noise power retains a non-zero component at zero temperature, acting as $S^Q(\omega) \sim |\omega|^3$. With zero frequency, it follows that $S^Q(0) = \kappa_0 T^3$.

The relation (6.1) applies to a one-dimensional chiral edge channel with linear dispersion and is also consistent with standard scattering theory [3]. Since Eq. (2.16) is an operator identity, this identity, and consequently Eq. (6.1), is valid for any edge channel, including the output channel specifically. By substituting the charge output noise power $S_{\text{out}}(\omega)$ from Eq. (2.21) into (6.1), we derive the output heat noise power as follows:

$$S^Q_{\text{out}}(\omega) = \frac{R_q^2}{4\pi} \int d\omega_1 \sum_{p=\text{in,oc}} |T_p(\omega_1)|^2 S_p(\omega_1) \times \sum_{k=\text{in,oc}} |T_k(\omega - \omega_1)|^2 S_k(\omega - \omega_1). \tag{6.3}$$

For the remainder of this subsection, our attention will be on the zero frequency heat noise power $S^Q_{\text{out}}(0)$. As a first check, we examine the global equilibrium scenario, where $T_{\text{in}} = T_{\text{oc}} = T$. Under these conditions, with the equilibrium charge noise power (2.22) and utilizing the sum rule (2.23), Eq. (6.3) simplifies to

$$S^Q_{\text{out}}(0) = \frac{1}{4\pi} \int d\omega_1 \frac{\omega_1}{1 - e^{-\omega_1/T}} \frac{-\omega_1}{1 - e^{\omega_1/T}} = \kappa_0 T^3, \tag{6.4}$$

which gives the expected equilibrium heat noise form as indicated in [2]. Note that we first took the limit $\omega \to 0$ so that Eq. (6.4) vanishes in the zero-temperature limit. The physical interpretation of (6.4) is simply that for a global uniform temperature T, the RC time of the OC is completely irrelevant for the heat noise, since the particle distributions are always at the same temperature, regardless of how long the particles dwell in the OC.

To this end, we consider the same two configurations as in the previous Chap. 2. For configuration (i), we take the limit of a very cold OC, $T_{\text{oc}} \to 0$, and insert the zero-temperature limit for $p = \text{oc}$ into the output heat noise power (6.3). After some algebraic manipulations, we arrive at the integral expression

$$S^Q_{\text{out}}(0) = \frac{T_{\text{in}}^3}{2\pi} \int_0^\infty dz \frac{e^z z^2 \left((\tau_C T_{\text{in}})^2 z^2 + 1\right) - (\tau_C T_{\text{in}})^2 z^4}{(e^z - 1)^2 \left((\tau_C T_{\text{in}})^2 z^2 + 1\right)^2}. \tag{6.5}$$

In the limits $\tau_C T_{\text{in}} \ll 1$ and $\tau_C T_{\text{in}} \gg 1$, we find that Eq. (6.5) can be analytically approximated as

$$\frac{S^Q_{\text{out}}(0)}{S^Q_{\text{in}}(0)} \simeq \begin{cases} 1 + \left(\frac{72\zeta(5)}{\pi^2} - \frac{8\pi^2}{5}\right)(\tau_C T_{\text{in}})^2 & \tau_C T_{\text{in}} \ll 1, \\ \frac{3}{4\pi} \frac{1}{\tau_C T_{\text{in}}} & \tau_C T_{\text{in}} \gg 1, \end{cases} \tag{6.6}$$

where $\zeta(z)$ is the Riemann zeta-function (with $\zeta(5) \approx 1.04$) and $S^Q_{\text{in}}(0) = \kappa_0 T^3_{\text{in}}$ is the zero frequency input heat noise power. From Eq. (6.6), we see that the heat noise power also reveals that for $\tau_C T_{\text{in}} \ll 1$, the OC cannot cool the input channel.

For configuration (ii), we instead take $T_{\text{in}} \to 0$, and keep T_{oc} finite. Thus, by inserting the zero temperature limit for $p = \text{in}$ into Eq. (6.3), we find the following.

$$S^Q_{\text{out}}(0) = \frac{T^3_{\text{oc}}}{2\pi} \int_0^\infty dz \frac{(\tau_C T_{\text{oc}})^2 z^4 \left(e^z \left((\tau_C T_{\text{oc}})^2 z^2 + 1\right) - 1\right)}{(e^z - 1)^2 \left((\tau_C T_{\text{oc}})^2 z^2 + 1\right)^2} \tag{6.7}$$

In the two limits $\tau_C T_{\text{oc}} \ll 1$ and $\tau_C T_{\text{oc}} \gg 1$, this integral can be evaluated analytically as

$$\frac{S^Q_{\text{out}}(0)}{S^Q_{\text{oc}}(0)} \simeq \begin{cases} 72 \frac{T^2_{\text{oc}} \tau^2_C}{\pi^2} \zeta(5) & \tau_C T_{\text{oc}} \ll 1, \\ 1 - \frac{3}{4\pi} \frac{1}{\tau_C T_{\text{oc}}} & \tau_C T_{\text{oc}} \gg 1, \end{cases} \tag{6.8}$$

where $S^Q_{\text{oc}}(0) = \kappa_0 T^3_{\text{oc}}$ is the zero frequency OC heat noise power. We thus see that for $\tau_C T_{\text{oc}} \gg 1$, the OC cannot heat the input channel.

In this section, we have illustrated that the phenomenon of heat Coulomb blockade—wherein the Ohmic contact's (OC's) finite capacitance hinders the equilibration between the OC and the edge channel—is also evident in the zero frequency heat noise power $S^Q_{\text{out}}(0)$. The influence of τ_C on heat noise is directly derived from the charge-heat relations expressed in Eqs. (2.16) and (6.1), which are applicable to one-dimensional chiral transport at low energies. Note that the effect of τ_C on heat noise emerges because we exactly account for current fluctuations, in contrast to the approach taken in [4].

We emphasize that the results discussed in this subsection are applicable only under the assumption of negligible Ohmic contact (OC) temperature fluctuations, $\Delta T_{\text{oc}}(t) = 0$, effectively assuming an infinite heat capacity, $C_E[T_{\text{oc}}(t)] \propto \tau_E \to \infty$. However, it is known, as noted in [4], that heat current fluctuations are influenced by temperature fluctuations and a finite energy relaxation time. It is therefore natural to ask whether is possible to observe effects arising not only from the finite OC capacitance but also finite OC heat capacity. This is the question to which we turn next.

6.2 Heat Noise Due to Temperature Fluctuations

In this section, we analyze the impact of sizeable temperature fluctuations of the Ohmic contact (OC) temperature, expressed as $T_{\text{oc}}(t) = T_{\text{oc}} + \Delta T_{\text{oc}}(t)$. These fluctuations arise from the finite OC heat capacitance $C_E[T_{\text{oc}}(t)]$, as illustrated in the setup depicted in Fig. 1.1.

6.2.1 Heat Current Langevin Source and Output Heat Current Fluctuations

To incorporate Ohmic contact (OC) temperature fluctuations, we now make explicit use of the separation of time scales detailed in Eq. (2.8): This separation implies that OC temperature fluctuations occur on a time scale significantly longer than that of charge fluctuations. To supplement the equations governing charge dynamics, (2.14a) and (2.14b), we introduce an equation for the conservation of energy, U, in the OC:

$$\frac{dU(t)}{dt} = J_{\text{in}}(t) - J_{\text{out}}(t). \tag{6.9}$$

Here, $J_{\text{in}}(t)$ and $J_{\text{out}}(t)$ represent the temporal incoming and outgoing heat currents, respectively. To formulate a Langevin equation that encompasses fluctuations in the outgoing heat current as well, a dual averaging procedure is employed. Averaging with respect to charge fluctuations, signified by simple brackets $\langle \cdots \rangle$, and temperature fluctuations, indicated by $\langle \cdots \rangle_E$, the rate of change of U equates to the heating/cooling power $P_{\text{heat/cool}}$ exerted on the OC, as follows:

$$\left\langle\left\langle \frac{d}{dt} U \right\rangle\right\rangle_E = P_{\text{heat/cool}}. \tag{6.10}$$

This allows to self consistently determine the temperature of the OC. At this point potential loss of energy due to phonons could be added, but we refrain from doing so. Due to the finite heat capacitance of the OC, the output heat current acquires fluctuations on a very long timescale because of the large assumption τ_E. This implies that fluctuations in internal energy can be approximated within the Langevin framework as

$$\Delta U(t) \approx C_E[T_{\text{oc}}]\Delta T_{\text{oc}}(t), \tag{6.11}$$

keeping ΔT_{oc} to linear order and thus considering the heat capacitance $C_E[T_{\text{oc}}(t)] \approx C_E[T_{\text{oc}}]$ to be dependent only on the average OC temperature T_{oc}. Contributions of higher orders of $\Delta T_{\text{oc}}(t)$ are negligible for large τ_E.

This allows for the formulation of a Langevin-like equation for the output heat current fluctuations, derived from the operator relation (2.16) as:

$$\delta J_{\text{out}}(t) = \kappa T_{\text{oc}} \Delta T_{\text{oc}}(t) + \delta J_{\text{out}}^{\text{fast}}(t). \tag{6.12}$$

Here, $\delta J_{\text{out}}(t) \equiv J_{\text{out}}(t) - \langle\langle J_{\text{out}}(t)\rangle\rangle_E$ represents the output heat current fluctuations, and

$$\delta J_{\text{out}}^{\text{fast}}(t) \equiv \frac{R_q}{2}\left[(\delta I_{\text{out}}(t))^2 - \langle(\delta I_{\text{out}}(t))^2\rangle\right] \tag{6.13}$$

serves as a semi-classical source of heat current fluctuations originating from the charge current fluctuations.

The noise power corresponding to $\delta J_{\text{out}}^{\text{fast}}(t)$ is given by the previously calculated heat noise power of the OC subject only to charge fluctuations (6.3), i.e.,

$$\langle\langle \delta J_{\text{out}}^{\text{fast}}(\omega)\delta J_{\text{out}}^{\text{fast}}(\omega')\rangle\rangle_E = 2\pi\delta(\omega+\omega')S_{\text{out}}^{Q}(\omega). \tag{6.14}$$

This identity connects the Langevin approaches for charge and heat currents and guarantees the consistency of our approach. However, unlike traditional Langevin theory, the source $\delta J_{\text{out}}^{\text{fast}}$ correlates with the incoming heat current fluctuations, leading to the relation:

$$\langle\langle \delta J_{\text{in}}(\omega)\delta J_{\text{out}}^{\text{fast}}(\omega')\rangle\rangle_E = 2\pi\delta(\omega+\omega')S_{\text{io}}^{Q}(\omega). \tag{6.15}$$

We obtain this cross-correlated noise power in a similar way as the output noise Eq. (6.3) and it can be written as

$$S_{\text{io}}^{Q}(\omega) = \frac{R_q^2}{4\pi}\int d\omega_1 |T_{\text{in}}(\omega_1)|^2 S_{\text{in}}(\omega_1)S_{\text{in}}(\omega-\omega_1). \tag{6.16}$$

In the zero frequency limit, this expression reduces to

$$S_{\text{io}}^{Q}(0) = \frac{T_{\text{in}}^3}{4\pi}\int_{-\infty}^{\infty} dz \frac{z^2}{(e^z-1)(1-e^{-z})}\frac{1}{((\tau_C T_{\text{in}})^2 z^2+1)}, \tag{6.17}$$

with the asymptotic values

$$\frac{S_{\text{io}}^{Q}(0)}{S_{\text{in}}^{Q}(0)} \simeq \begin{cases} 1-\frac{4\pi^2}{5}(\tau_C T_{\text{in}})^2 & \tau_C T_{\text{in}} \ll 1, \\ \frac{3}{2\pi}\frac{1}{\tau_C T_{\text{in}}} & \tau_C T_{\text{in}} \gg 1. \end{cases} \tag{6.18}$$

We now turn our attention to the first term in Eq. (6.12), which represents the linear response coefficient of the output heat current to a slight variation in the Ohmic contact (OC) temperature, $T_{\text{oc}}(t)$. Consequently, we define:

$$\kappa T_{\text{oc}} \equiv \frac{\partial}{\partial T_{\text{oc}}}\langle\langle J_{\text{out}}\rangle\rangle_E\,. \tag{6.19}$$

To compute this quantity, we differentiate the output heat current (2.26) with respect to T_{oc} and by using Eq. (2.27), we find

$$\kappa T_{\text{oc}} = \kappa_0 T_{\text{oc}} + \frac{1}{4\pi T_{\text{oc}}\tau_C^2} + \frac{1}{4\tau_C} - \frac{\psi'\left(\frac{1}{2\pi T_{\text{oc}}\tau_C}\right)}{8\pi^2 T_{\text{oc}}^2\tau_C^3}. \tag{6.20}$$

Here, $\psi'(z)$ refers to the derivative of the digamma function $\psi(z)$, which was defined just after Eq. (2.28). Intriguingly, Eq. (6.20) reveals that the coefficient κ tends towards distinct asymptotic values based on the dimensionless parameter $\tau_C T_{\text{oc}}$:

$$\frac{\kappa}{\kappa_0} = \begin{cases} \frac{4\pi^2}{5}\tau_C^2 T_{\mathrm{oc}}^2 & \text{if } \tau_C T_{\mathrm{oc}} \ll 1, \\ 1 - \frac{3}{2\pi}\frac{1}{\tau_C T_{\mathrm{oc}}} & \text{if } \tau_C T_{\mathrm{oc}} \gg 1. \end{cases} \tag{6.21}$$

In particular, we see that in the heat Coulomb blockade regime, a new time-scale emerges:

$$\tilde{\tau}_E \equiv \frac{C_E[T_{\mathrm{oc}}]}{\kappa T_{\mathrm{oc}}} \sim \frac{\tau_E}{T_{\mathrm{oc}}^2 \tau_C^2} \gg \tau_E, \quad \tau_C T_{\mathrm{oc}} \ll 1. \tag{6.22}$$

This time scale is much larger than the energy relaxation time τ_E, under the assumption of separation of time scales (2.8). Equation (6.22) implies that in the Coulomb heat blockade regime, $\tau_C T_{\mathrm{oc}} \ll 1$, the OC temperature fluctuations, induced by the finite $C_E[T_{\mathrm{oc}}]$, are affected, as shown explicitly in Eq. (6.32). Furthermore, the explicit dependence of $\tilde{\tau}_E$ on κ in Eq. (6.21) shows that the finite heat capacity of the OC suppresses also the single edge channel heat conductance, which without the OC is simply $\kappa = \kappa_0$.

Proceeding in a manner similar to the calculation of charge current noise discussed in Chap. 2, we now evaluate the output heat current noise resulting from temperature fluctuations. Utilizing the Fourier transform, Eqs. (6.9) and (6.12) are solved to determine the fluctuations in the output heat current $\delta J_{\mathrm{out}}(\omega)$. The solution emerges as a linear combination of the input heat currents and OC heat currents:

$$\delta J_{\mathrm{out}}(\omega) = \mathcal{T}_{\mathrm{in}}^Q(\omega)\delta J_{\mathrm{in}}(\omega) + \mathcal{T}_{\mathrm{out}}^Q(\omega)\delta J_{\mathrm{out}}^{\mathrm{fast}}(\omega), \tag{6.23a}$$

$$\mathcal{T}_{\mathrm{in}}^Q(\omega) = 1 - \mathcal{T}_{\mathrm{out}}^Q(\omega) = \left[1 - i\omega\tilde{\tau}_E\right]^{-1}, \tag{6.23b}$$

with $\tilde{\tau}_E = C_E[T_{\mathrm{oc}}]/(\kappa T_{\mathrm{oc}})$ now reflecting the influence of heat Coulomb blockade, according to Eq. (6.22). Contrary to their charge counterparts in (2.18b), the linear coefficients $\mathcal{T}_p^Q(\omega)$ for $p \in \{\mathrm{in}, \mathrm{out}\}$ in Eq. (6.23a) should not be interpreted as scattering amplitudes for bosonic heat fluctuations. Their similarity arises because we are using linearized equations for the internal energy. By combining Eqs. (6.23a) and (6.23b), we derive the power of heat noise in scenarios involving temperature fluctuations. To clearly differentiate this from the heat noise power computed without temperature fluctuations—as discussed in Sect. 6.1—we introduce a specific notation for the heat noise power affected by both temperature and voltage fluctuations as:

$$\chi_{\mathrm{out}}^Q(\omega) = 2\pi\delta(\omega + \omega')\left\langle\!\left\langle \delta J_{\mathrm{out}}(\omega) J_{\mathrm{out}}(\omega')\right\rangle\!\right\rangle_E. \tag{6.24}$$

In other words, $\chi_{\mathrm{out}}^Q(\omega)$ is the heat noise power with the constraint that frequencies are of the order of the inverse energy relaxation time $\omega\tau_E \sim 1$ and τ_E respect the separation of time scales, that is, $\tilde{\tau}_E^{-1} \ll T_{\mathrm{in}}, T_{\mathrm{oc}}$. This justifies the use of the Langevin approach for the heat current, however, enforces that the heat noise power of the fast fluctuations $S_{\mathrm{in}}^Q(\omega)$, $S_{\mathrm{io}}^Q(\omega)$ and $S_{\mathrm{out}}^Q(\omega)$ enters $\chi_{\mathrm{out}}^Q(\omega)$ only in the zero frequency limit.

Solving Eqs. (6.23a) and (6.23b), we then find

$$\chi^{Q}_{\text{out}}(\omega) = \sum_{p=\text{in,out}} |\mathcal{T}^{Q}_{p}(\omega)|^2 \, S^{Q}_{p}(0). \tag{6.25}$$

where $|\mathcal{T}^{Q}_{\text{in}}(\omega)|^2 = (1+\omega^2\tilde{\tau}_E^2)^{-1} = 1 - |\mathcal{T}^{Q}_{\text{out}}(\omega)|^2$ and $\langle \delta J_p(\omega)\delta J_p(\omega')\rangle_E = 2\pi\delta(\omega+\omega')S^{Q}_{p}(\omega)$. The structural similarity between Eq. (6.25) and the formula for the output charge noise power in Eq. (2.21) is notable, upon exchanging $\tau_C \leftrightarrow \tilde{\tau}_E$ and $\mathcal{T}_p \leftrightarrow \mathcal{T}^{Q}_{p}$. Interestingly, the cross noise power $S^{Q}_{\text{io}}(\omega)$ does not contribute to this formula, as $\mathcal{T}^{Q}_{\text{in}}(\omega)\mathcal{T}^{Q}_{\text{out}}(-\omega) = -\mathcal{T}^{Q}_{\text{out}}(\omega)\mathcal{T}^{Q}_{\text{in}}(-\omega)$, leading to an exact cancellation in the final expression of the noise power $\chi^{Q}_{\text{out}}(\omega)$.

Similar to the dynamics of charge, there exists a sum rule for the heat dynamics:

$$\sum_{p=\text{in,out}} |\mathcal{T}^{Q}_{p}(\omega)|^2 = 1, \tag{6.26}$$

which here reflects energy conservation. In the zero frequency limit, we see that

$$\chi^{Q}_{\text{out}}(0) = S^{Q}_{\text{in}}(0), \tag{6.27}$$

so that the heat fluctuations averaged over very long time scales do not "see" effects of the OC heat capacity.

6.2.2 *Temperature Fluctuations*

We can now address the effects of temperature fluctuations, brought into focus by the introduction of the novel energy relaxation time (6.22) in the Coulomb blockade regime, through the calculation of the temperature-temperature correlation function. This function is defined as:

$$\langle\langle \Delta T_{\text{oc}}(\omega)\Delta T_{\text{oc}}(\omega')\rangle\rangle_E = 2\pi\delta(\omega+\omega')S^{T}_{\text{oc}}(\omega). \tag{6.28}$$

We obtain this function by solving Eqs. (6.9), (6.11), and (6.23a) and obtain

$$S^{T}_{\text{oc}}(\omega) = \frac{S^{Q}_{\text{in}}(0) - 2S^{Q}_{\text{io}}(0) + S^{Q}_{\text{out}}(0)}{\kappa^2 T^2_{\text{oc}}\left(1+\omega^2\tilde{\tau}_E^2\right)}. \tag{6.29}$$

For example, for $T_{\text{in}} = T_{\text{oc}} = T$, we find

$$S^{T}_{\text{oc}}(\omega) = \frac{2T}{\kappa\left(1+\omega^2\tilde{\tau}_E^2\right)}, \tag{6.30}$$

which, quite surprisingly, takes the same functional form independently of the asymptotic limits we consider. This is related to the fact that in equilibrium the following relation holds.

$$\begin{aligned} &S^Q_{\text{in}}(0) - 2S^Q_{\text{io}}(0) + S^Q_{\text{out}}(0) \\ &\quad = R_q^2 \int \frac{d\omega'}{2\pi} \left(1 - |\mathcal{T}_{\text{in}}(\omega')|^2\right) S(\omega')S(-\omega') \\ &\quad\quad = T^2 \partial_T R_q \int \frac{d\omega'}{2\pi} |\mathcal{T}_{\text{oc}}(\omega')|^2 S(\omega) = 2\kappa T^3, \end{aligned} \tag{6.31}$$

Note that the thermal conductance κ given by Eq. (6.21) and thus the energy relaxation time $\tilde{\tau}_E$ is different depending on whether the system is in the heat Coulomb blockade regime or not. In the time-domain, Eq. (6.30) corresponds to an exponential suppression of the temperature correlations

$$\langle\langle \Delta T_{\text{oc}}(t) \Delta T_{\text{oc}}(0) \rangle\rangle_E = \frac{T^2}{C_E[T]} e^{-|t|/\tilde{\tau}_E}. \tag{6.32}$$

Note that the amplitude of the temperature-temperature correlation function is the universal, expected result [13], but its decay rate, $\tilde{\tau}_E$, is in fact sensitive to the charge relaxation time (or the OC capacity C) via Eq. (6.22).

6.3 Multi-channel Heat Dynamics

For the N-channel heat dynamics, Eqs. (6.9) and (6.12) generalize to

$$\frac{dU(t)}{dt} = \sum_{m=1}^{N} \left[J_{\text{in},m}(t) - J_{\text{out},m}(t) \right], \tag{6.33}$$

$$\delta J_{\text{out},m}(t) = \kappa_N T_{\text{oc}} \Delta T_{\text{oc}}(t) + \delta J^{\text{fast}}_{\text{out},m}(t). \tag{6.34}$$

Similarly to the one-channel scenario (6.2), the fast source in the Langevin-type equation (6.33) can have non-trivial cross correlations with the incoming heat current fluctuations

$$\left\langle\left\langle \delta J_{\text{in,m}}(\omega) \delta J^{\text{fast}}_{\text{out},m'}(\omega') \right\rangle\right\rangle_E = 2\pi \delta(\omega + \omega') S^{Q,N}_{\text{io}}(0). \tag{6.35}$$

We assume right from the start, that the cross heat noise power of the source is not dependent on the specific channel. This is true if $\mathcal{T}_{\text{in,m}} = \mathcal{T}_{\text{in}}$. Similar to (6.17) we find the asymptotic values of this cross heat noise

$$\frac{S_{\rm io}^{Q,N}(0)}{\kappa_0 T_{\rm in}^3} \approx \begin{cases} \frac{1}{N^2} - \frac{4\pi^2}{5N^2}(\tau_C T_{\rm in})^2 & \tau_{C,N} T_{\rm in} \ll 1, \\ \frac{3}{2\pi N^2}\frac{1}{\tau_{C,N} T_{\rm in}} & \tau_C T_{\rm in} \gg 1. \end{cases} \tag{6.36}$$

Additionally the source-source correlation function shows non trivial correlations of the form

$$\langle\langle \delta J_{\rm out,m}^{\rm fast}(\omega)\delta J_{\rm out,m'}^{\rm fast}(\omega')\rangle\rangle_E = 2\pi\delta(\omega+\omega')\left(S_{\rm out}^Q(0) + S_{\rm oo}^Q(0)\right), \tag{6.37}$$

where we defined the cross heat noise power

$$S_{\rm oo}^Q(0) = \frac{R_q}{4\pi}\int d\omega' S_{\rm out,m,m'}(\omega')S_{\rm out,m,m'}(-\omega'). \tag{6.38}$$

which is related to the charge cross noise power (2.37). Note that the contribution from $S_{\rm oo}^Q(0)$ vanishes in equilibrium due to the structure of the charge cross noise power (2.38). We again assumed no spatial dependence of the auto $S_{{\rm out},m}^Q(0) = S_{\rm out}^Q(0)$ and cross $S_{{\rm oo},mm'}^Q(0) = S_{\rm oo}^Q(0)$ heat noise powers.

Here, $U(t)$ is the internal OC energy and $J_{{\rm in},m}(t)$ and $J_{{\rm out},m}^{\rm fast}(t)$ are the incoming and outgoing heat currents on branch m, respectively. The linear response coefficient of the output heat flux in channel m, when N channels are attached, is given as

$$\kappa_N T_{\rm oc} \equiv \partial_{T_{\rm oc}} \langle\langle J_{{\rm out},m}\rangle\rangle_E. \tag{6.39}$$

For which we find

$$\frac{\kappa_N}{\kappa_0} = \begin{cases} 1 - \frac{3}{2\pi}\frac{1}{N\tau_{C,N} T_{\rm oc}} & \tau_{C,N} T_{\rm oc} \gg 1, \\ 1 - \frac{1}{N} + \frac{4\pi^2}{5N}\tau_{C,N}^2 T_{\rm oc}^2 & \tau_{C,N} T_{\rm oc} \ll 1. \end{cases} \tag{6.40}$$

Important differences between the N-channel (6.40) and the single channel (6.20) thermal conductances is that τ_C is replaced by $\tau_{C,N} = \tau_C/N$ and that in the N-channel setup, the output heat noise contribution from the OC capacitance is distributed equally between the output channels. In other words, we have

$$\kappa_N - \kappa_0 = \frac{1}{N}(\kappa_1 - \kappa_0)\Big|_{\tau_C \to \tau_{C,N}}, \tag{6.41}$$

where κ_1 is identified with the single channel contribution (6.19). With the thermal conductance (6.40), the solution of Eqs. (6.33) and (6.34) is given by

$$\delta J_{{\rm out},m}(\omega) = \sum_{p={\rm in,out}} \left(\boldsymbol{T}_p^Q(\omega)\right)_{mn} \delta J_{p,n}(\omega), \tag{6.42a}$$

$$\left(\boldsymbol{T}_{\rm out}^Q(\omega)\right)_{mn} \equiv \delta_{mn} - N^{-1}\left[1 - i\omega\tilde{\tau}_{E,N}\right]^{-1} \tag{6.42b}$$

$$\left(\boldsymbol{T}_{\text{in}}^{Q}(\omega)\right)_{mn} \equiv N^{-1}\left[1 - i\omega\tilde{\tau}_{E,N}\right]^{-1}, \tag{6.42c}$$

which allows us to compute the heat noise power $\chi^{Q}_{\text{out},m}(\omega)$ as

$$\chi^{Q}_{\text{out},m}(\omega) = \frac{S^{Q}_{\text{in}}(0) + S^{Q}_{\text{out}}(0)\left[N - 1 + N\omega^2\tilde{\tau}^2_{E,N}\right]}{N(1+\omega^2\tilde{\tau}^2_{E,N})} - \frac{S^{Q}_{\text{oo}}(0)}{1+\omega^2\tilde{\tau}^2_{E,N}}. \tag{6.43}$$

Importantly, the total energy relaxation time of the OC is $\tilde{\tau}_{E,N} \equiv \tilde{\tau}_E/N$ where $\tilde{\tau}_E = C_E[T_{\text{oc}}]/(\kappa_N T_{\text{oc}})$ for the multi-channel case. This modified relaxation time account for the charge-heat coupling. We have also used that when the temperature fluctuations are taken into account, the input and OC heat noise powers are given by $S^{Q}_{p}(0)$, for $p \in \{\text{in}, \text{out}\}$, paralleling the discussion on the time-scale separation in Sect. 6.2. Note that the sum rule (6.26) in the N channel system still holds, reflecting energy conservation. This is due to the cancelation of $S^{Q}_{\text{oo}}(0)$ in equilibrium.

With the same choices, we also compute the cross-correlation heat noise powers, defined for $m \neq n$ as

$$\langle \delta J_{\text{out},m}(\omega)\delta J_{\text{out},n}(\omega')\rangle \equiv 2\pi\delta(\omega+\omega')\chi_{\text{out},m,n}(\omega), \tag{6.44}$$

and obtain

$$\chi^{Q}_{\text{out},m,n}(\omega) = \frac{S^{Q}_{\text{in}}(0) - S^{Q}_{\text{out}}(0)}{N(1+\omega^2\tilde{\tau}^2_{E,N})} + \frac{S^{Q}_{\text{oo}}(0)\omega^2\tilde{\tau}^2_{E,N}}{1+\omega^2\tilde{\tau}^2_{E,N}}. \tag{6.45}$$

A relation similar to that between auto- and cross-correlation charge noise powers can be found also for the heat noise powers due to the structural similarity of the equations. This relation reads

$$\chi^{Q}_{\text{out},m,n}(\omega) = \chi^{Q}_{\text{out},m}(\omega) + S^{Q}_{\text{oo}}(0) - S^{Q}_{\text{out}}(0). \tag{6.46}$$

6.3.1 Temperature Fluctuations

The solution of the Langevin equation (6.42a) allows us to compute the temperature fluctuations in the N channel setup

$$\begin{aligned}\left\langle \Delta T_{\text{oc}}(\omega)\Delta T_{\text{oc}}(\omega')\right\rangle &= 2\pi\delta(\omega+\omega')\\ &\times \frac{1}{\kappa_N^2 T_{\text{oc}}^2}\frac{S^{Q}_{\text{in}}(0) - 2\,N S^{Q}_{\text{io}}(0) + N S^{Q}_{\text{oo}}(0) + S^{Q}_{\text{out}}(0)}{N(1+\omega^2\tilde{\tau}^2_{E,N})}.\end{aligned} \tag{6.47}$$

Let us again consider equal temperatures $T_{\text{in}} = T_{\text{oc}} = T$, which makes the cross noise power $S^{Q}_{\text{oo}}(0)$ vanish from the expression of (6.47). Next, using the asymptotic form of $S^{Q}_{\text{io}}(0)$, given by (6.36), we obtain

$$\left\langle \Delta T_{\rm oc}(\omega)\Delta T_{\rm oc}(\omega')\right\rangle = 2\pi\delta(\omega+\omega')\frac{1}{\kappa_N}\frac{2T}{N(1+\omega^2\tilde{\tau}_{E,N}^2)}, \tag{6.48}$$

thanks to the remarkable cancellation, similar to Eq. (6.31):

$$S_{\rm in}^{Q}(0) - 2\,N S_{\rm io}^{Q}(0) + S_{\rm out}^{Q}(0) = 2\kappa_N T^3, \tag{6.49}$$

in equilibrium at equal temperatures. The temperature-temperature correlation function (6.47) can also be calculated for non-equilibrium scenarios.

6.4 Non-gaussian Counting Statistics from Temperature Fluctuations

6.4.1 Fourth Cumulant

For negligible temperature fluctuations, it can be shown [5] that the edge channel OC system in Fig. 6.1 can be described by a quadratic bosonized model, cf. Eq. (5.2). This feature implies that the full counting statistics (FCS) [6, 7] of the output charge

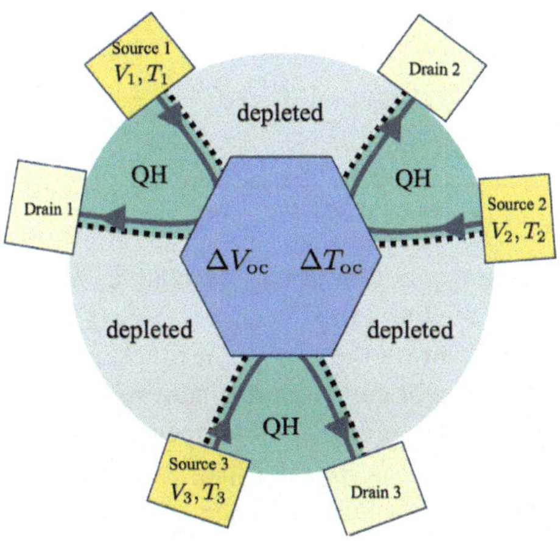

Fig. 6.1 Repetition of Fig. 2.3. Schematics of an Ohmic contact (OC, in blue) connected to multiple, chiral edge channels in the quantum Hall (QH) regime. Here, three pairs, $N = 3$, of edge channels are depicted. Pairs of incoming/outgoing edge channels appear on the boundaries (black, dotted lines) between QH bulk regions (turquoise regions, filling factor ν) and depleted regions (gray regions). Each pair is referred to as a "branch" in the main text. All input channels are fed from large metallic contacts (in yellow) characterized by input voltages, $V_{\rm in}$, and input temperatures, $T_{\rm in}$, for $m = 1, \ldots, N$. Fluctuations carried by the input edge channels generate voltage, $\Delta V_{\rm oc}$, and temperature, $\Delta T_{\rm oc}$, fluctuations. In turn, these quantities produce charge and heat fluctuations in the outgoing channels, which can be detected in the drain contacts (in light yellow)

and heat currents only contain first and second cumulants, that is, the statistics are Gaussian.

In this section, we demonstrate a novel and complementary effect from sizeable OC temperature fluctuations, namely non-Gaussian FCS. To this end, we consider a cross section of the output channel m far away from the OC. We define the net charge that passes this cross-section in the long time interval $\tau_{\text{obs}} \sim \tau_E$ as

$$\delta q(\tau_{\text{obs}}) \equiv \int_{-\tau_{\text{obs}}/2}^{\tau_{\text{obs}}/2} dt\, \delta I_{\text{out,m}}(t), \tag{6.50}$$

where we, for notational ease, denote $\delta I_{\text{out},m}(t) = \delta I(t)$ in the remainder of this section. Without temperature fluctuations, the system is, as mentioned above, Gaussian, and quantities like the fourth cumulant of the transferred charge

$$\left\langle (\delta q(\tau_{\text{obs}}))^4 \right\rangle_c \equiv \left\langle (\delta q(\tau_{\text{obs}}))^4 \right\rangle - 3 \left(\left\langle (\delta q(\tau_{\text{obs}}))^2 \right\rangle \right)^2, \tag{6.51}$$

vanish due to Wick's theorem. However, in the presence of temperature fluctuations, we now show that the double averaging procedure produces additional corrections $\Delta \langle\langle \delta I(t_1)\delta I(t_2)\delta I(t_3)\delta I(t_4)\rangle\rangle_E$, to Eq. (6.51). These can be considered as "noise of noise" contributions, similar to the "cascade corrections" considered in [8–12].

By explicitly considering the temperature dependence of the two point functions, we can write the correction to the fourth cumulant as

$$\begin{aligned}\Delta \langle\langle \delta I(t_1)\delta I(t_2)\delta I(t_3)\delta I(t_4)\rangle\rangle_E &= 3 \int dt dt' \\ &\times \frac{\delta \langle \delta I(t_1)\delta I(t_2)\rangle\, \delta \langle \delta I(t_3)\delta I(t_4)\rangle}{\delta T_{\text{oc}}(t)\delta T_{\text{oc}}(t')} \left\langle \Delta T_{\text{oc}}(t)\Delta T_{\text{oc}}(t')\right\rangle_E, \end{aligned}\tag{6.52}$$

where $\delta \cdots / \delta T_{\text{oc}}(t)$ denotes a functional derivative with respect to the time-dependent temperature of the OC $T_{\text{oc}}(t) = T_{\text{oc}} + \Delta T_{\text{oc}}(t)$ and the factor of 3 comes from the different ways of pairing the two-point function.

The outgoing current-current correlation function, $\langle \delta I(t_1)\delta I(t_2)\rangle$, loses its time translation invariance due to the presence of temperature fluctuations on a long time scale. By Fourier transforming $S_{\text{out},m}(\omega)$ in Eq. (2.35), we see that the time difference $t_d = t_1 - t_2 \sim \tau_C$ governs the dynamics of the correlation function on short time scales:

$$\langle \delta I(t_1)\delta I(t_2)\rangle = \int_{-\infty}^{\infty} \frac{d\omega}{2\pi} e^{-i\omega(t_1-t_2)} S_{\text{out},m}(\omega, T_{\text{oc}}(f(t_1,t_2))), \tag{6.53}$$

where $f(t_1, t_2)$ is, for the moment, an arbitrary function of the times t_1, t_2. We now assume that the time dependence of the temperature $T_{\text{oc}}(t) \approx T_{\text{oc}}(t_s)$ depends very slowly on time, i.e., on the scale $t_s = (t_1 + t_2)/2 \sim \tau_E$. Since we consider characteristic frequencies $\omega\tau_C \ll 1$, we can evaluate the integral in Eq. (6.53) only over t_d and approximate $T_{\rm oc}(t_s) \approx T_{\rm oc}$. These approximations imply further that

$$\frac{\delta \langle \delta I(t_1)\delta I(t_2)\rangle}{\delta T_{\rm oc}(t)} \approx \frac{d S_{{\rm out},m}(0)}{d T_{\rm oc}} \delta(t - t_s), \tag{6.54}$$

where $S_{{\rm out},m}(0)$ is given in Eq. (2.36). Plugging the approximation (6.54) into the correction term (6.52), and using the expression for the temperature correlation function in Eq. (6.47), we find that the fourth cumulant (6.51) does not vanish, but instead evaluates to

$$\left\langle (\delta q(\tau_{\rm obs}))^4 \right\rangle_c = 3\left(\frac{N-1}{R_q N}\right)^2 \int_{-\tau_{\rm obs}/2}^{\tau_{\rm obs}/2} dt \int_{-\tau_{\rm obs}/2}^{\tau_{\rm obs}/2} dt' \left\langle \Delta T_{\rm oc}(t)\Delta T_{\rm oc}(t')\right\rangle_E \tag{6.55}$$

with

$$\int_{-\tau_{\rm obs}/2}^{\tau_{\rm obs}/2} dt \int_{-\tau_{\rm obs}/2}^{\tau_{\rm obs}/2} dt' \left\langle \Delta T_{\rm oc}(t)\Delta T_{\rm oc}(t')\right\rangle_E = \tau_{\rm obs} \int_{-\infty}^{\infty} dy \frac{\sin\left(\frac{y}{2}\right)}{\pi y} S^T_{\rm oc}\left(\frac{y}{\tau_{\rm obs}}\right), \tag{6.56}$$

where $S^T_{\rm oc}(\omega)$ is defined in Eq. (6.47). This yields for the fourth cumulant of transmitted charge

$$\left\langle (\delta q(\tau_{\rm obs}))^4 \right\rangle_c = 3\tau_{\rm obs}\left(\frac{N-1}{R_q N}\right)^2 \frac{1 - e^{-\tau_{\rm obs}/(2\tilde{\tau}_E)}}{N\kappa_N^2 T_{\rm oc}^2} \times \left(S^Q_{\rm in}(0) - 2N S^Q_{\rm io}(0) + N S^Q_{00}(0) + S^Q_{\rm out}(0)\right). \tag{6.57}$$

For the special case $N = 1$, the OC temperature fluctuations do not produce a nonzero fourth cumulant, even for finite $\tilde{\tau}_{E,N}$. The reason is that for a single connected channel, the output charge current and zero frequency charge noise power do not depend on $T_{\rm oc}(t)$ due to charge conservation, see Eqs. (2.19) and (2.24). The fourth cumulant is therefore, in this case, unaffected by the OC temperature fluctuations and vanishes. Note that the equilibration process, i.e. $\tau_{\rm e-e} \ll \tau_E$ inside of the OC is crucial. A mesoscopic capacitor has also a finite level spacing or Thouless time, but the scattering is completely elastic. This leads to energy being conserved at every instance of time, hence one requirement to detect a fourth cumulant is to have a local non-equilibrium distribution function instead of a global equilibrium one. This distinction between mesoscopic capacitors and metallic granulas does not appear if one discussed only average heat fluxes, where both models behave qualitatively equivalent.

6.4.2 Third Cumulant in the Presence of Finite Voltage Bias

Until now, we focused on the case of no voltage bias in the system, i.e., $V_{\text{in},m} = 0$ so that there is no average potential on the OC, $V_{\text{oc}} = 0$. To obtain a finite correction, to the third cumulant $\left\langle(\delta q(\tau_{\text{obs}}))^3\right\rangle_c \equiv \left\langle(\delta q(\tau_{\text{obs}}))^3\right\rangle$, we will show that a finite V_{oc} is needed. Detecting the third cumulant may be more experimentally feasible than the fourth cumulant, which is why we perform this additional calculation at finite V_{oc}.

If we voltage bias a single input channel, (see Fig. 2.3), e.g., by taking $V_{\text{in},m} = V\delta_{m,1}$, the energy dynamics of the system is no longer fully described by heat currents. Instead we must in this equation take:

$$J_{\text{in},m}(t) \to I^E_{\text{in},m}(t) = J_{\text{in},m}(t) + V\delta_{m,1}I_{\text{in},m}(t), \tag{6.58a}$$

$$J_{\text{out},m}(t) \to I^E_{\text{out},m}(t) = J_{\text{out},m}(t) + V_{\text{oc}}I_{\text{out},m}(t), \tag{6.58b}$$

where from Kirchoff's voltage law, we have that $V_{\text{oc}} = V/N$. From conservation of the average energy currents $\sum_m I^E_{\text{in},m} = \sum_m I^E_{\text{out},m}$, we find that the average OC temperature in the presence of the bias becomes

$$\kappa_m T^2_{\text{oc}} = \kappa_0 T^2_{\text{in}} + \frac{N-1}{N^2}\frac{V^2}{R_q}. \tag{6.59}$$

With the substitutions (6.58), the energy conservation Eq. (6.33) determining the OC temperature fluctuations then generalizes to

$$\begin{aligned} \frac{d\Delta U(t)}{dt} &\approx C_E[T_{\text{oc}}]\frac{d\Delta T_{\text{oc}}(t)}{dt} \\ &= \sum_{m=1}^{N}\left[\delta J_{\text{in},m}(t) - \delta J^{\text{tot}}_{\text{out},m}(t) + V\left(\delta I_{\text{in},1}(t)\delta_{m,1} - N^{-1}\delta I_{\text{out},m}(t)\right)\right], \end{aligned} \tag{6.60}$$

which thus contains a crucial term that is linear in the charge-current fluctuations. We now show that this modification results in a correction to the third cumulant of the emitted charge at finite bias. This correction, denoted ΔC_3, takes the form

$$\Delta C_3 \equiv \prod_{n=1}^{3}\int_{-\tau_{\text{obs}}/2}^{\tau_{\text{obs}}/2} dt_n \left\langle\langle\delta I(t_1)\delta I(t_2)\delta I(t_3)\rangle\right\rangle_E,$$

where we remind that $\delta I(t)$ is short-hand for $\delta I_{\text{out},m}(t)$. Just as for the fourth cumulant, the third cumulant vanishes in the absence of temperature fluctuations due to the Gaussian nature of the charge fluctuations, leaving the correction ΔC_3 as the only non-vanishing component.

With the same approach as for the fourth cumulant above, we write the leading order correction term as

$$\Delta \mathcal{C}_3 \approx 3 \prod_{n=1}^{3} \int_{-\tau_{\text{obs}}/2}^{\tau_{\text{obs}}/2} dt_n \int_{-\infty}^{\infty} dt \frac{\delta \langle \delta I(t_1) \delta I(t_2) \rangle}{\delta T_{\text{oc}}(t)} \langle \langle \Delta T_{\text{oc}}(t) \delta I(t_3) \rangle \rangle_E. \quad (6.61)$$

To compute these time integrals, we start by solving the energy conservation equation (6.60) and then combine it with the heat Langevin equation (6.42a). We apply the separation of fast and slow time scales (6.54). Finally, by using the identity for the charge-current noises at zero frequency, we arrive at

$$\mathcal{C}_3 \approx \Delta \mathcal{C}_3 = -3\tau_{\text{obs}} \left(\frac{N-1}{R_q N} \right)^2 \times \left(1 - e^{-\tau_{\text{obs}}/(2\tilde{\tau}_{E,N})} \right) \frac{V T_{\text{in}}}{N \kappa_m T_{\text{oc}}}, \quad (6.62)$$

with the average OC electrical potential $V_{\text{oc}} = V/N$. We see that this expression is indeed finite only when the voltage bias is $V \neq 0$ and $N \neq 1$. The requirement of $N > 1$ is, as for the fourth cumulant, necessary to depend on $T_{\text{oc}}(t)$ for the emitted charge current and zero-frequency charge current noise.

The third cumulant (6.62) was obtained in [12] for the particular case of $N = 2$ and a symmetric cavity in the regime $\tau_{C,N} T_{\text{oc}} \gg 1$. Our calculation thus generalizes the third cumulant to generic values of $\tau_{C,N} T_{\text{oc}}$. We further note that for $N > 2$, neither the third nor the fourth cumulant coincides with the corresponding cumulants in [12] even in the regime $\tau_{C,N} T_{\text{oc}} \gg 1$. This happens due to different definitions of the transmitted charge in the chiral and non-chiral conduction channels for $N > 2$. We finally remark that a finite voltage bias would modify the fourth cumulant.

References

1. Idrisov EG, Levkivskyi IP, Sukhorukov EV (2022) Thermal drag effect in quantum Hall circuits. Phys Rev B 106:L121405. https://link.aps.org/doi/10.1103/PhysRevB.106.L121405
2. Averin DV, Pekola JP (2010) Violation of the fluctuation-dissipation theorem in time-dependent mesoscopic heat transport. Phys Rev Lett 104:220601. https://doi.org/10.1103/PhysRevLett.104.220601
3. Blanter Ya. M, Büttiker M (2000) Shot noise in mesoscopic conductors. Phys Rep 336:1–166. https://doi.org/10.1016/S0370-1573(99)00123-4
4. Dashti N, Misiorny M, Samuelsson P, Splettstoesser J (2018) Probing charge- and heat-current noise by frequency-dependent fluctuations in temperature and potential. Phys Rev Appl 10:024007. https://doi.org/10.1103/PhysRevApplied.10.024007
5. Slobodeniuk AO, Levkivskyi IP, Sukhorukov EV (2013) Equilibration of quantum Hall edge states by an Ohmic contact. Phys Rev B 88:165307. https://link.aps.org/doi/10.1103/PhysRevB.88.165307
6. Levitov LS, Lesovik GB (1993) Charge distribution in quantum shot noise. Sov J Exp Theor Phys Lett 58:230. https://ui.adsabs.harvard.edu/abs/1993JETPL..58..230L/abstract
7. Levitov LS, Lee H, Lesovik GB (1996) Electron counting statistics and coherent states of electric current. J Math Phys 37:4845–4866. https://doi.org/10.1063/1.531672
8. Nagaev KE (2002) Cascade Boltzmann-Langevin approach to higher-order current correlations in diffusive metal contacts. Phys Rev B 66:075334. https://doi.org/10.1103/PhysRevB.66.075334

9. Nagaev KE, Samuelsson P, Pilgram S (2002) Cascade approach to current fluctuations in a chaotic cavity. Phys Rev B 66:195318. https://doi.org/10.1103/PhysRevB.66.195318
10. Pilgram S, Nagaev KE, Büttiker M (2004) Frequency-dependent third cumulant of current in diffusive conductors. Phys Rev B 70:045304. https://doi.org/10.1103/PhysRevB.70.045304
11. van den Berg TL, Brange F, Samuelsson P (2015) Energy and temperature fluctuations in the single electron box. New J Phys 17:075012. https://doi.org/10.1088/1367-2630/17/7/075012
12. Pilgram S, Jordan AN, Sukhorukov EV, Büttiker M (2003) Stochastic path integral formulation of full counting statistics. Phys Rev Lett 90:206801. https://link.aps.org/doi/10.1103/PhysRevLett.90.206801
13. Pekola JP, Karimi B (2021) Colloquium: quantum heat transport in condensed matter systems. Rev Mod Phys 93:041001. https://doi.org/10.1103/RevModPhys.93.041001

Chapter 7
Outlook

Abstract This chapter presents a collection of open questions and speculative ideas that emerged during the main investigations. We explore non-chiral transmission lines, analyzing charge fluctuations in a diffusive system, and discovering a novel mathematical identity related to eigenvalue products of a finite difference operator. A discrete Green's function approach is developed to resolve boundary condition challenges in non-chiral transport. Extending the study to fractional quantum Hall edge states, we derive modified heat flux and Lorenz numbers, revealing potential deviations from the Wiedemann-Franz law. Additionally, we investigate the impact of temperature fluctuations on the nonequilibrium distribution function, showing how they lead to non-Gaussian full-counting statistics (FCS) and potentially affect tunneling properties. Finally, we discuss the limitations of the tunneling Hamiltonian approach beyond leading order, particularly in systems with feedback loops, such as mesoscopic capacitors. These insights open new avenues for studying heat transport, correlated states, and non-equilibrium effects in mesoscopic systems.

7.1 Non-chiral Transmission Line

This section extends the concept of the transmission line to purely diffusive systems. Consider a transmission line in which each node is connected by an edge state entering from both the left $\delta I_{\text{in,L}}$ and right $\delta I_{\text{in,R}}$, as well as an edge state exiting from both the left $\delta I_{\text{out},L}$ and right $\delta I_{\text{out},R}$, see Fig. 7.1 for a schematic setup.[1] We consider a TL composed of N Ohmic contacts (OCs) and assume that the boundary currents exhibit no voltage fluctuations,[2] i.e. $\delta I_{\text{in,R}}(N+1,t) = \delta I_{\text{res,R}}(t)$ and $\delta I_{\text{in,L}}(0,t) = \delta I_{\text{res,L}}(t)$. Furthermore, we specify that the incoming current of one node corresponds to the outgoing current of the adjacent node, establishing that $\delta I_{\text{in,R}}(j,t) = \delta I_{\text{out},L}(j+1,t)$

[1] This resembles a tower of multipliers, discussed in Chap. 5, where an OC that sits in the "loop is repeated N times.

[2] They may originate from a large reservoir.

F. Stäbler, *Dissipation, Interactions, and Heat Transport in Quantum Hall Devices*, Springer Theses, https://doi.org/10.1007/978-3-031-90671-8_7

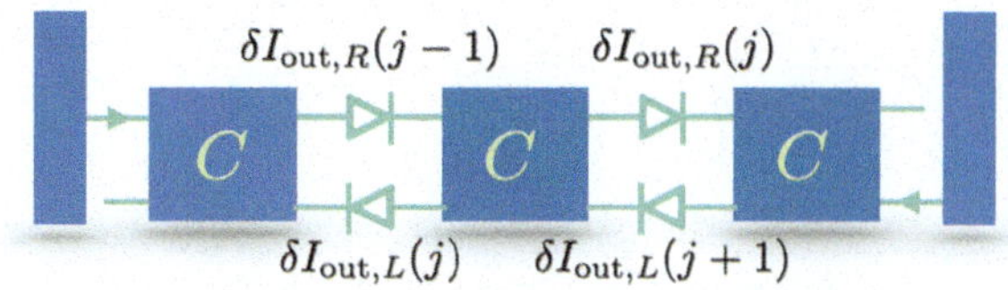

Fig. 7.1 The labelling of currents of in the non-chiral TL schematically shown for three OCs

and $\delta I_{\text{in,L}}(j,t) = \delta I_{\text{out},R}(j-1,t)$. The current associated with each Ohmic contact (OC) obeys a Langevin equation:

$$\delta I_{\text{out},\alpha}(j,t) = \frac{1}{\tau_C}\Delta Q(j,t) + \delta I_{\text{oc},\alpha}(j,t). \tag{7.1}$$

Kirchhoff's law is then expressed as:

$$\begin{aligned}(2 - i\omega\tau_C)\Delta Q(j) - \Delta Q(j+1) - \Delta Q(j-1) \\ = \tau_C\left(\delta I_{\text{oc,L}}(j+1) - \delta I_{\text{oc,L}}(j) + \delta I_{\text{oc,R}}(j-1) - \delta I_{\text{oc,R}}(j)\right).\end{aligned} \tag{7.2}$$

Note that the spectrum of the collective mode ΔQ is given by the eigenvalues of the finite difference operator in (7.2), subject to Dirichlet boundary conditions, since we demand that $\Delta Q(0,t) = \Delta Q(N+1,t) = 0$. This gives a set of Eigenvalues $1 < n < N$:

$$i\omega_n = \frac{2}{\tau_C}\left(1 - \cos\left(\frac{n\pi}{N+1}\right)\right). \tag{7.3}$$

7.1.1 A New Identity?

Manipulating these expressions led to a new type of mathematical identity, which to my knowledge is previously unknown.[3] The product over all of the eigenfrequencies represents the determinant of the finite difference operator and is given by

$$D(N) = \prod_{n=1}^{N} \omega_n = \tau_C^{-N}\left(e^{\frac{i\pi}{1+N}}; e^{\frac{i\pi}{1+N}}\right)_N^2, \tag{7.4}$$

where $(a;q)_n = \prod_{k=0}^{n-1}\left(1 - aq^k\right)$ is the q-analogue of the Pochhammer function. One can check numerically that $i^N D(N) = N+1$ for integer values of N. However the finite difference method presented here is equivalent to the N dimensional matrix problem

$$(M_N)_{nm} = 2\delta_{nm} + \delta_{nm-1} + \delta_{nm+1} \tag{7.5}$$

[3] Counterexamples or a more rigorous proof is highly appreciated.

Using Jacobi's method of expressing the determinant of M_N in terms of the determinants of the smaller matrices M_{N-1} and M_{N-2} yields the recursion equation

$$\det M_N = 2 \det M_{N-1} - \det M_{N-2}, \tag{7.6}$$

which is solved by the characteristic equation

$$(r - 1)^2 = 0, \tag{7.7}$$

and has a double root at $x_0 = 1$, thus the solution is then given by

$$\det M_N = (k_1 + k_2 n)(x_0)^n, \tag{7.8}$$

with the initial conditions $\det M_0 = 1$ and $\det M_1 = 2$.

This gives

$$k_1 = 1, \quad (1 + k_2) = 2 \rightarrow k_2 = 1, \tag{7.9}$$

and thus

$$\det M_N = N + 1 \stackrel{?}{=} i^N \tau_C^N D(N). \tag{7.10}$$

7.1.2 Discrete Green's Function Method

The most important difference between the chiral transmission line, presented in Chap. 3 and the present case is that the charge conservation Eq. (7.2) cannot be solved by a Fourier transform without modifications due to the special spectrum (7.3), which makes the collective mode susceptible to the boundary conditions imposed on the transmission line. This is reflected in an unphysical divergence of the left/right heat flux in the limit of $\tau_C \rightarrow 0$, upon false use of the standard Fourier transform. We proceed instead using a discrete Green's function method expressed in terms of the eigenfunctions of the finite difference operator

$$\varphi_k(n) = \sqrt{\frac{2}{N+1}} \sin\left(\frac{\pi n k}{N+1}\right). \tag{7.11}$$

The discrete Green's function of our equation of motion may be defined as

$$(2 - i\omega\tau_C)G(n, n') - G(n+1, n') - G(n-1, n') = \delta_{nn'}, \tag{7.12}$$

which is solved by the following function

$$G(n, n') = \sum_k \frac{1}{\lambda_k} \varphi_k(n) \varphi_k(n'). \tag{7.13}$$

Let us write down the equation with the source term

$$\mathcal{D}Q(n) = f(n), \tag{7.14}$$

where $\mathcal{D}$ is the difference operator we solved before. The solution is given in terms of the discrete convolution

$$Q(n) = \sum_{n'} G(n, n')f(n'), \tag{7.15}$$

The solution of the equation of motion

$$\begin{aligned}(2 - i\omega\tau_C)\Delta Q(j) - \Delta Q(j+1) - \Delta Q(j-1)& \\ = \tau_C \left(\delta I_{\text{oc,L}}(j+1) - \delta I_{\text{oc,L}}(j) + \delta I_{\text{oc,R}}(j-1) - \delta I_{\text{oc,R}}(j)\right),&\end{aligned} \tag{7.16}$$

is given by

$$\begin{aligned}&\Delta Q(n)\\ &= \tau_C \sum_{n'} G(n, n') \left(\delta I_{\text{oc,L}}(j+1) - \delta I_{\text{oc,L}}(j) + \delta I_{\text{oc,R}}(j-1) - \delta I_{\text{oc,R}}(j)\right).\end{aligned} \tag{7.17}$$

The Langevin equation for the right-going current at position n is given by

$$\delta I_R(n) = \frac{\Delta Q(n)}{\tau_C} + \delta I_{\text{oc,R}}(n), \tag{7.18}$$

which formally gives (we ignore already all correlation functions of the left, since only correlation functions of the current to the right are correlated.)

$$\begin{aligned}\langle \delta I_R(n,\omega)\delta I_R(n,-\omega)\rangle &= \left\langle \delta I_{\text{oc,R}}(n,\omega)\delta I_{\text{oc,R}}(n,-\omega)\right\rangle \\ &+ \sum_{n'} G(n,n',-\omega)\left\langle \delta I_{\text{oc,R}}(n,\omega)\left(\delta I_{\text{oc,R}}(n'-1,-\omega) - \delta I_{\text{oc,R}}(n',-\omega)\right)\right\rangle \\ &+ \sum_{n'} G(n,n',\omega)\left\langle \delta I_{\text{oc,R}}(n,-\omega)\left(\delta I_{\text{oc,R}}(n'-1,\omega) - \delta I_{\text{oc,R}}(n',\omega)\right)\right\rangle \\ &+ \sum_{n',n''} G(n,n',\omega)G(n,n'',-\omega) \\ &\times \left\langle\left(\delta I_{\text{oc,R}}(n'-1,\omega) - \delta I_{\text{oc,R}}(n',\omega)\right)\left(\delta I_{\text{oc,R}}(n''-1,-\omega) - \delta I_{\text{oc,R}}(n'',-\omega)\right)\right\rangle \\ &+ \sum_{n',n''} G(n,n',\omega)G(n,n'',-\omega) \\ &\times \left\langle\left(\delta I_{\text{oc,L}}(n'+1,\omega) - \delta I_{\text{oc,L}}(n',\omega)\right)\left(\delta I_{\text{oc,L}}(n''+1,-\omega) - \delta I_{\text{oc,L}}(n'',-\omega)\right)\right\rangle,\end{aligned} \tag{7.19}$$

This yields the noise power

$$
\begin{aligned}
S_{\text{out},R}(n,\omega) = S_{\text{oc,R}}(n,\omega)(1 &+ G(n,n+1,-\omega) - G(n,n,-\omega) \\
&+ G(n,n+1,\omega) - G(n,n,\omega) \\
&+ 2\sum_{n'} 2G(n,n',\omega)G(n,n',-\omega) \\
- G(n,n',\omega)G(n,n'-1,-\omega) &- G(n,n',\omega)G(n,n'+1,-\omega)),
\end{aligned} \tag{7.20}
$$

where the last line can be simplified, since

$$
\begin{aligned}
&2G(n,n',\omega) - G(n,n'-1,\omega) - G(n,n'+1,\omega) \\
&\quad = \sum_k \frac{1}{\lambda_k} 4\sin^2\left(\frac{k\pi}{2(N+1)}\right)\sin\left(\frac{kn\pi}{N+1}\right)\sin\left(\frac{kn'\pi}{N+1}\right).
\end{aligned} \tag{7.21}
$$

In the limit of strong interactions $\tau_C \to 0$ everything simplifies nicely. A useful identity is

$$
\sum_{k=1}^{N} \frac{\sin\left(\frac{kn\pi}{N+1}\right)\sin\left(\frac{kn'\pi}{N+1}\right)}{1-\cos\left(\frac{k\pi}{N+1}\right)} = n(N+1-n'), \quad 1 \le n \le n' \le N. \tag{7.22}
$$

In the strong-interaction limit, we are left with

$$
S_{\text{out},R}(n,\omega) = S_{\text{oc,R}}(n,\omega)\left(1 + \frac{2}{N+1}\sum_k \frac{\sin\left(\frac{kn\pi}{N+1}\right)\sin\left(\frac{k(1+n)\pi}{1+N}\right)}{\left(1-\cos\left(\frac{k\pi}{1+N}\right)\right)}\right), \tag{7.23}
$$

where we used

$$
\sum_{n'} \underbrace{\left(2G(n,n',\omega) - G(n,n'-1,\omega) - G(n,n'+1,\omega)\right)}_{\delta_{nn'}} G(n,n',\omega) = G_{n,n}. \tag{7.24}
$$

This gives the final result.

$$
\begin{aligned}
S_{\text{out},R}(n,\omega) &= S_{\text{oc,R}}(\omega) \\
&\times\left(1 + \frac{2}{N+1}\left(n(N-n) - \frac{1}{2}n(N+1-n)\right)\right) \\
&= S_{\text{oc,R}}\left(1 + \frac{2(N-n)n}{1+N}\right).
\end{aligned} \tag{7.25}
$$

For inspiration on the discrete Green function approach, we may check [1]. The noise power of the right-going current in Eq. (7.25) translates to a heat flux carried to the right given by

$$J_{\text{out},R}(n) = J_q(1 + \frac{2(N-n)n}{1+N}). \tag{7.26}$$

The heat flux to the left is described by the same equation, indicating that in equilibrium, the heat flux per cross section is zero. The equation for right-going heat flux (7.26) aligns with the solutions obtained from the equations of motion for $N = 2, 3, 4$ and solving them by hand. Notably, the heat flux carried by the edge varies with the position n where it is measured and tends toward infinity as $N \to \infty$. This observation lays the groundwork for intriguing approaches to modeling interacting diffusive one-dimensional systems both in equilibrium and out of it. Future research could explore topics such as the Wiedemann-Franz law, Fourier law, anomalous correlation functions, non-chiral two-dimensional networks, and other fascinating phenomena.

7.1.3 Solving for the Zero Frequency Green's Function

Let us start from the differential equation for the Green function directly

$$(-i\omega\tau_C + 2)G(n, n') - G(n+1, n') - G(n-1, n') = \delta_{nn'} \tag{7.27}$$

Let us first consider the case of $\tau_C = 0$. In that case

$$2G(n, n') - G(n+1, n') - G(n-1, n') = 0, \tag{7.28}$$

if $n \neq n'$. Rearranging this equation gives

$$\begin{aligned} G(n, n') - G(n-1, n') \\ &= G(n-1, n') - G(n-2, n') \\ &= \cdots = G(1, n'), \end{aligned} \tag{7.29}$$

since due to the Dirichlet boundary conditions $G(0, n') = 0$. This means that the ansatz

$$G(n, n') = nG(1, n'), \tag{7.30}$$

solves Eq. (7.31) up to a constant factor. We can write another type of equation using the symmetry of n, n'

where the action of the finite difference operator $\mathbf{F}_n(\omega)$ is defined by

$$\tilde{\mathbf{F}}_n(\omega) f_n = -i\omega f_n + \frac{1}{\tau_C}\left(\nu_{\mathrm{L}}\left(f_n - f_{n+1}\right) + \nu_{\mathrm{R}}\left(f_n - f_{n-1}\right)\right), \tag{7.54}$$

The eigenvalues of this finite difference operator are given by the eigenvalues of the corresponding tridiagonal Toeplitz matrix:

$$\lambda_k = -i\omega + \frac{1}{\tau_c}\left(\nu_{\mathrm{R}} + \nu_{\mathrm{L}} - 2\sqrt{\nu_{\mathrm{R}}\nu_{\mathrm{L}}}\cos\left(\frac{k\pi}{N+1}\right)\right). \tag{7.55}$$

This allows us to compute the Green's function for different boundary conditions using the spectral decomposition and, the heat flux in every part of the system, and the Lorenz number. Note that the noise power of the sources will be affected by the modified electrical conductances $R_q \to R_q/\nu_{\mathrm{R/L}}$.

An analysis of all combinations of boundary conditions and potentially the non-equilibrium scenarios goes beyond the scope of this letter and will be considered elsewhere. As an example we will compute the Green's function with open boundary conditions for general filling factors ν_L and ν_R in the heat Coulomb blockade regime:

$$G(n, m, 0) = \begin{cases} \frac{(1-\gamma^n)(1-\gamma^{N+1-m})}{\nu_L(1-\gamma)(1-\gamma^{N+1})} & \text{for } n \leq m \\ \\ \frac{\gamma^{N+1}(1-\gamma^{-m})(1-\gamma^{-(N+1-n)})}{\nu_L(1-\gamma)(1-\gamma^{N+1})} & \text{for } n \geq m \end{cases} \tag{7.56}$$

where we defined $\gamma = \nu_R/\nu_L$. For $\nu_L = \nu_R = 1$, the Green's function (7.56) reduces to the one computed in Eq. (7.36). In equilibrium, i.e., $T_n = T$, the heat flux transported to the left/right at a given point n is given by

$$J_n^\sigma = \left(1 + 2\nu_\sigma G(n, n+\sigma, 0)\right) J_q, \tag{7.57}$$

with the Green's function (7.56) found earlier and $\sigma = \pm 1$ for L/R. Note that the heat flux only depends on the ratio of filling factors γ, which is evident after evaluating (7.57) explicitly. The Lorenz number in the aforementioned tunneling experiment would be

$$L(n, N, \sigma) = 3L_0\left(1 - \frac{2}{3 + 2\nu_\sigma G(n, n+\sigma, 0)}\right). \tag{7.58}$$

As a specific example, consider the reconstructed fill factor $\nu = 2/3$ that features an upstream mode with $\nu_R = 1$ and a downstream mode with $\nu_L = 1/3$ for $n \to 1$ and $N \to 2$. In that case, we expect to find

$$J_n^\sigma = 1 + \frac{3\left(1 + 3^N - 3^{N-n} - 3^n\right)}{3^{N+1} - 1} J_q \overset{\substack{n\to 1\\ N\to 2}}{=} \frac{19}{13} J_q, \tag{7.59}$$

and

$$L(n, N, \sigma) = 3 + \frac{2\,3^n\left(3^{N+1}-1\right)}{3^N - 4\,3^{N+n} + 9^n} \overset{\substack{n\to 1\\ N\to 2}}{=} \frac{19}{15}L_0, \tag{7.60}$$

which is a heat flux or Lorenz number that is slightly lower than for $\nu_L = 1$ and $\nu_R = 1$.

7.2 Non-equilibrium Distribution Function Due to Temperature Fluctuations

The discovery that temperature fluctuations result in non-Gaussian FCS, as detailed in Sect. 6.4, could have significant implications for the fermionic distribution function $f(\varepsilon)$ in an outgoing channel of the Ohmic contact (OC).

$$f(\epsilon) = \int dt e^{i\epsilon t} K(t), \tag{7.61a}$$

$$K(t) = \langle \psi^\dagger(t)\psi(0)\rangle, \tag{7.61b}$$

The fermionic correlation function $K(t)$ can be expressed in terms of the second C_2 and fourth cumulant C_4 (and other cumulants if present)

$$K(t) = \frac{1}{2\pi a}\left\langle e^{-i\phi(t)}e^{i\phi(0)}\right\rangle = \frac{1}{2\pi a}e^{-\sum_n C_n}, \tag{7.62}$$

where C_2 is the gaussian contribution and C_4 is related to the fourth cumulant of transmitted charge computed in Sect. 6.4. Special care has to be taken when computing tunneling current or heat flux, similar to Chap. 5 in the presence of these higher order cumulants. Are they maybe suppressed? This direction is still an open question.

7.3 Tunneling Hamiltonian Approach Beyond Leading Order

In Chap. 5, the tunneling Hamiltonian approach was utilized to perturbatively treat tunneling to leading order. However, extending the study of these looped states beyond the leading order presents a challenge. The tunneling Hamiltonian, as described in (5.25c), is expressed using bosonic fields, under the assumption that these fields satisfy the scattering problem at the Ohmic contact (OC).

$$\gamma\psi_+(x_0,t)\psi_-(x_0,t) = \frac{\gamma}{2\pi a}e^{-i(\phi_+(x_0,t)-\phi_-(x_0,t))}. \tag{7.63}$$

Beyond the leading order in tunneling, the fields ϕ_+ and ϕ_- depend on the perturbative tunneling strength γ, a consequence of the self-looping nature of the edge states in the multiplier setup. It is an open question how this dependence alters the conventional S-matrix and thus the tunneling Hamiltonian approach and its potential impact on the outcomes beyond leading order discussed in Chap. 5. Various scenarios could be envisioned where this would be relevant, for example one can imagine a mesoscopic capacitor, like the one in Fig. 4.8, where one allows tunneling from the incoming state directly to the outgoing state. The loop inside of the capacitor represents a similar feedback and would most certainly affect the electronic correlation function in equilibrium.

References

1. Chung F, Yau S-T (2000) Discrete green's functions. J Comb Theory, Ser A 91:191–214. https://doi.org/10.1006/jcta.2000.3094
2. Stäbler F, Sukhorukov E (2023) Mesoscopic heat multiplier and fractionalizer. Phys Rev B 108(23):235405. https://doi.org/10.1103/PhysRevB.108.235405
3. Sukhorukov EV (2016) Scattering theory approach to bosonization of non-equilibrium mesoscopic systems. Physica E 77:191–198. https://doi.org/10.1016/j.physe.2015.11.018
4. Kiselev MN (2023) Generalized Wiedemann-Franz law in a two-site charge Kondo circuit: Lorenz ratio as a manifestation of the orthogonality catastrophe. Phys Rev B 108(8):L081108. https://doi.org/10.1103/PhysRevB.108.L081108

Chapter 8
Summary

Abstract This summary chapter reviews the key findings of the thesis, which explores Ohmic contacts in quantum Hall edge systems across four core projects. It outlines the development of theoretical models addressing heat transport, dissipation, and Coulomb blockade effects, including the "missing heat" paradox and nonlocal energy transfer. The chapter highlights how interactions and edge self-looping shape thermal behavior and fluctuations, tying together the broader insights gained into quantum correlations and transport in low-dimensional systems.

This thesis represents a comprehensive compilation of research undertaken during my doctoral studies, focusing on Ohmic contacts (OCs) within quantum hall edge systems. The work is divided into four distinct projects, each exploring different facets of OCs and their implications for experiments in quantum Hall (QH) systems. These projects, now published as individual papers, are further developed in this thesis, providing more detailed insights and making the findings more accessible.

We begin with an in-depth exploration of the OC as an element and theoretical model in contemporary quantum Hall experiments. The discussion includes the theoretical model of the OC connected to QH edge states, and returns established results associated with the heat Coulomb blockade effect in single- and multichannel setups.

Further, the thesis introduces the transmission line (TL) model as a framework to understand dissipation phenomena and potential nonlocal effects, such as diffusive currents, emerging from disorder within QH edge states at the microscopic level. We provide an introduction to the "missing heat" paradox, where experimental observations suggest an unaccounted for loss of energy in QH edge states. Our analysis shows that dissipation alone cannot explain the observed energy discrepancy. The thesis also establishes a link between equilibrium heat flux within the edge state and a sum rule that ensures equilibrium is achieved.

Next, we discuss how nonlocal couplings in an equilibrium drift-diffusion system can circumvent the aforementioned sum rule, allowing for heat transfer in other parts of the system. This results in a QH edge state carrying heat in excess of the equilibrium heat flux quantum, attributed to additional correlations from strong interactions.

We then shift to an OC featuring self-looping edge states, drawing parallels with the non-local TL systems discussed earlier. Beyond contributions from free fermions,

F. Stäbler, *Dissipation, Interactions, and Heat Transport in Quantum Hall Devices*, Springer Theses, https://doi.org/10.1007/978-3-031-90671-8_8

these loops carry additional correlations enhancing the heat flux due to the OC's self-interaction. Despite the edge state appearing hotter than the device's equilibrium temperature, energy extraction via a thermometer is shown to be impossible, highlighting the conservation of energy principle. However, unexpected behaviors are revealed when probing these states with a tunneling quantum point contact, manifesting the correlated nature through interaction-modified correlation functions, which affect electrical and thermal tunneling conductances and scaling dimensions.

The thesis culminates in an exploration of the heat Coulomb blockade's signature in heat noise power and temperature fluctuations of the OC stemming from its finite heat capacitance. A Langevin-like approach is adopted to address thermal conductance and energy relaxation times within the Coulomb blockade regime. The investigation of the temperature-temperature correlation function reveals its universal equilibrium form, uninfluenced by the heat Coulomb blockade. Moreover, the discovery that temperature fluctuations induce non-Gaussian full-counting statistics (FCS) of transmitted charge downstream of the device underscores the link between thermal fluctuations and quantum transport phenomena.

In conclusion, the thesis not only sheds light on the complex dynamics and correlations within OCs and QH edge states, but also lays the groundwork for future research in this domain, presenting a collection of ideas and unfinished projects as intellectual contributions for subsequent investigations and interested readers. In a broader context, this thesis significantly enhances our understanding of correlations and heat transport within interacting low-dimensional systems, paving the way for advancements in electronics miniaturization, precision metrology, and the realization of quantum information technologies.

Appendix A
TL—Dissipation

A.1 Heat Flux of the Symmetric Mode

For the symmetric mode, we find

$$
\begin{aligned}
J^s &= R_q \left\langle \left(\delta I^s_{\text{out}}(j,t)\right)^2 \right\rangle \\
&= R_q \sum_{k,q} \int \frac{d\omega d\omega'}{4\pi^2} e^{i\frac{2\pi(k+q)}{N}j - i(\omega+\omega')t} \left\langle \delta I^s_{out}(k,\omega) \delta I^s_{out}(q,\omega') \right\rangle \\
&= R_q \int \frac{d\omega dk}{4\pi^2} \frac{\omega^2 \tau_C^2 S^s_{\text{oc}}(\omega)}{2 + \omega^2\tau_C^2 - 2\cos(k) - 2\omega\tau_C \sin(k)}, \qquad \text{(A.1)}
\end{aligned}
$$

where we took the large N limit $\frac{1}{N}\sum_{k=0}^{N-1} \to \int_0^{2\pi} \frac{dk}{2\pi}$. The sources are delta-correlated for momenta and frequencies in the following way

$$
\left\langle \delta I^s_{out}(k,\omega) \delta I^s_{out}(q,\omega') \right\rangle = \frac{\delta_{k,-q}}{N} 2\pi \delta(\omega+\omega') S^s_{\text{oc}}(\omega), \qquad \text{(A.2a)}
$$

$$
S^s_{\text{oc}}(\omega) = \frac{1}{2} S(\omega), \qquad \text{(A.2b)}
$$

with the noise power given by the standard expression (2.22). The k-integral can be mapped onto the unit circle by the transformation $z = e^{ik}$ or equivalently $k = -i\log(z)$ and $dk = dz/(iz)$. The integral has one pole inside the unit circle at $z^* = (1 - i\omega\tau_C)^{-1}$. Thus, we find

$$
\begin{aligned}
J^s &= R_q \int \frac{d\omega dz}{4\pi^2} \frac{i\omega^2\tau_C^2 S^s_{\text{oc}}(\omega)}{1 + i\omega\tau_C - z(2+\tau_C^2\omega^2) + z^2(1 - i\omega\tau_C)} \\
&= R_q \int \frac{d\omega}{2\pi} S^s_{\text{oc}}(\omega) = J_q. \qquad \text{(A.3)}
\end{aligned}
$$

F. Stäbler, *Dissipation, Interactions, and Heat Transport in Quantum Hall Devices*, Springer Theses, https://doi.org/10.1007/978-3-031-90671-8

A.1.1 Heat Flux of the Dipole Mode

We repeat the same procedure for the dipole mode J^a

$$J^a = R_q \left\langle \left(\delta I^a_{\text{out}(j,t)}\right)^2 \right\rangle$$
$$= R_q \sum_{k,q} \int \frac{d\omega d\omega'}{4\pi^2} e^{i\frac{2\pi(k+q)}{N}j - i(\omega+\omega')t} \left\langle \delta I^a_{out}(k,\omega)\delta I^a_{out}(q,\omega') \right\rangle$$
$$= R_q \int \frac{d\omega dk}{4\pi^2} \frac{S_\perp(\omega) + S^a_{\text{oc}}(\omega)\left(\Gamma^2 + \omega^2\tau_C^2\right)}{2 + \Gamma(2+\Gamma) + \omega^2\tau_C^2 - 2(1+\Gamma)\cos(k) - 2\omega\tau_C\sin(k)}, \quad \text{(A.4)}$$

which can be rewritten in using the z variable. The integral has one pole inside the contour at $z^* = (1 + \Gamma - i\omega\tau_C)^{-1}$. This gives

$$J^a = R_q \int \frac{d\omega dz}{4\pi^2} \frac{i\left(S_\perp(\omega) + S^a_{\text{oc}}(\omega)\left(\Gamma^2 + \omega^2\tau_C^2\right)\right)}{(1 - z + \Gamma + i\omega\tau_C)(1 - z(1 + \Gamma - i\omega\tau_C))}$$
$$= R_q \int \frac{d\omega}{2\pi} \frac{S_\perp(\omega) + S^a_{\text{oc}}(\omega)(\Gamma^2 + \omega^2\tau_C^2)}{(2\Gamma + \Gamma^2 + \omega^2\tau_C^2)}$$
$$= \int \frac{d\omega}{4\pi} S(\omega) = J_q, \quad \text{(A.5)}$$

where $S^a_{\text{oc}}(\omega) = \frac{1}{2}S(\omega)$ and $S_\perp(\omega) = \Gamma S(\omega)$ in equilibrium, i.e. for equal temperatures.

A.2 IR-Theory from Exact Theory

In real experimental situations the temperature is usually small and only the low-energy part of the spectrum will be excited. We will expand the dispersion relation up to the point where the heat integrals converge giving a correct low-energy description for the chiral edge. All nonlinear contributions will be small in the continuum limit, but are nonetheless important for the understanding of the low-energy theory. This allows us to make three simplifications: (i) the main contribution of the integral comes from small momenta. The cutoff can be taken to infinity. We will estimate the error of the tails of the integral at a later point; (ii) The parameter ξ is small in the continuum limit and can be treated perturbatively; (iii) We rescale the capacitance of the system, introducing the velocity $v_q = \frac{\xi}{\tau_C}$. We assume that only the low-energy part of the spectrum is excited, i.e. $\omega \approx v_q k$ and we can introduce a small parameter $\tilde{k} = \omega/v_q - k$, which we can treat perturbatively.

A.2.1 *Low Energy Theory of the Charged Mode*

We restore the distance between the nodes ξ. The heat flux is given by

$$
\begin{aligned}
J^s &= \frac{R_q}{8\pi^2}\int d\omega \int_{-\pi/\xi}^{\pi/\xi} dk \frac{\xi\omega^2 S(\omega)}{(\omega+\omega^s(k))\,(\omega-\omega^s(k))} \\
&= \frac{R_q}{8\pi^2}\int d\omega \int_{-\pi/\xi}^{\pi/\xi} dk \frac{\xi\omega^2 \tau_C^2 S(\omega)}{(\omega\tau_C - k\xi)^2 + \frac{1}{3}k^3\xi^3\omega\tau_C - \frac{1}{12}k^4\xi^4} \\
&= \frac{R_q}{8\pi^2}\int d\omega \int_{-\pi/\xi}^{\pi/\xi} dk \frac{\xi\omega^2}{\left(\omega - v_q k\right)^2 + \frac{1}{3}k^3\xi^2 v_q\omega - \frac{1}{12}k^4\xi^2 v_q^2} S(\omega), \qquad \text{(A.6)}
\end{aligned}
$$

where we expanded the dispersion up to third order using the smallness of ξ, which gives $\omega^s(k) \approx k v_q - i\frac{k^2\xi v_q}{2} - \frac{k^3\xi^2 v_q}{6}$. For $\xi \to 0$ the integrand has a singularity $k^* = \omega/v_q$. Deviations from the singularity are small for small ξ, as required by the low energy approach. We can thus shift the integral by $\tilde{k} \to k - \omega/v_q$ and expand the cubic and quartic part to order $O(\tilde{k}^0)$.

$$
\begin{aligned}
J^s &= \frac{R_q}{8\pi^2}\int d\omega \int_{-\pi/\xi}^{\pi/\xi} dk \frac{\xi\omega^2 S(\omega)}{\left(\omega - v_q k\right)^2 + \frac{1}{3}k^3\xi^2 v_q\omega - \frac{1}{12}k^4\xi^2 v_q^2} \\
&\sim \frac{R_q}{8\pi^2}\int d\omega \int_{-\infty}^{\infty} d\tilde{k} \frac{\xi\omega^2 S(\omega)}{v_q^2\tilde{k}^2 + \frac{R_q}{4}\frac{\omega^4\xi^2}{v_q^2}} = \int \frac{d\omega}{4\pi} S(\omega) = J_q. \qquad \text{(A.7)}
\end{aligned}
$$

A.2.1.1 Corrections Due to a Finite Cut-Off

$$
\begin{aligned}
J^s &= \frac{R_q}{8\pi^2}\int d\omega \int_{-\pi/\xi}^{\pi/\xi} d\tilde{k} \frac{\xi\omega^2}{v_q^2\tilde{k}^2 + \frac{1}{4}\frac{\omega^4\xi^2}{v_q^2}} S(\omega) \\
&= R_q \int \frac{d\omega}{8\pi^2} 4\arctan\left(\frac{2\pi\omega^2}{v_q^2\xi^2}\right) S(\omega) \\
&\approx R_q \int \frac{d\omega}{8\pi^2}\left(2\pi - \frac{2\xi^2\omega^2}{\pi v_q^2}\right) S(\omega) = J_q - \frac{1}{\pi^2}J_4, \qquad \text{(A.8)}
\end{aligned}
$$

with $J_4 = \pi^3/30\ \tau_C^2 T^4$.

A.2.1.2 Corrections Due to Linear $\tilde{k}$ Term

$$J^s = \frac{R_q}{8\pi^2}\int d\omega \int_{-\infty}^{\infty} d\tilde{k} \frac{\xi\omega^2}{v_q^2\tilde{k}^2 + \frac{1}{4}\frac{\omega^4\xi^2}{v_q^2} + \frac{2\tilde{k}\xi^2\omega^3}{3v}} S(\omega)$$
$$= R_q \int \frac{d\omega}{4\pi} 3v_q \sqrt{\frac{1}{9v_q^2 - 4\xi^2\omega^2}} S(\omega)$$
$$\approx R_q \int \frac{d\omega}{8\pi^2}\left(2\pi + \frac{\xi^2\omega^2}{18\pi v_q^2}\right) S(\omega) = J_q + \frac{1}{36\pi^2} J_4. \quad \text{(A.9)}$$

A.2.2 *Low Energy Theory of the Neutral Mode*

The spectrum of the neutral mode is different. It contains now also the transverse resistor. To have a meaningful low-energy theory we want to analyze the heat flux in the limit of infinitesimal $\Gamma = R_q/R_\perp$. This additional term regularizes the integral already in the order $O(k^2)$ of the dispersion relation. The heat flux of the neutral mode is given by

$$J^a = \frac{R_q}{4\pi^2}\int d\omega \int_{-\infty}^{\infty} dk\xi \frac{S^a_{\mathrm{oc}}(\omega)(\Gamma^2 + \omega^2\tau_C^2) + S_\perp(\omega)}{(\omega\tau_C - k\xi)^2 + \Gamma^2 + k^2\xi^2\Gamma}$$
$$\approx \frac{R_q}{4\pi^2} > \int d\omega \int_{-\infty}^{\infty} dk \frac{\frac{\Gamma}{\xi}}{(\frac{\omega}{v_q} - k)^2 + \frac{\Gamma^2}{\xi^2}} S(\omega) = J_q, \quad \text{(A.10)}$$

we use the fact that $\Gamma \approx k\xi \ll 1$. This means that terms like $\Gamma\frac{\Gamma}{\xi}$, $k^2\Gamma$ and $\omega^2/v_q^2\xi$ will be of next leading order. The same integral as the last line of Eq. (A.10) can be also obtained by only taking into account the terms $\omega^a(k) \approx v_q k - i\frac{v_q\Gamma}{\xi}$. of the spectrum. Thus the integral is regularized already at the linear order of the expansion of the spectrum.

Appendix B
TL—Non-local Coupling

B.1 Cross-Capacitive Coupling

B.1.1 Strongly Interacting Limit $\lambda \to \frac{1}{2}$

In this limit the poles of Eq. (4.23) follow from the two polynomials

$$1 - z^2 - z^3 + z(1 + 2i\omega\tau_C) \stackrel{!}{=} 0 \tag{B.1a}$$

$$1 + z - z^3 - z^2(1 - 2i\omega\tau_C)) \stackrel{!}{=} 0, \tag{B.1b}$$

where $z \to e^{ik\xi}$. We refrain from writing the poles explicitly, but the structure of the poles can be seen in Fig. B.1a for the first polynomial and Fig. B.1b for the second polynomial in Eq. 4.23. Note the structural difference of the poles in quadrant (I) and (IV), belonging to the collective mode contribution at $k \to 0$ and the two poles in quadrant (II) and (III), which appear due to the nonlocal coupling and the approach $k \to \pi$ for $\omega \to 0$ indicating the closing of a gap of a second mode, which becomes soft sufficiently close to this point. Upon picking up the three poles inside the contour for $\omega > 0$ and $\omega < 0$ separately and assuming small temperatures, we find an intermediate result, which if integrated over frequencies directly yields Eq. (4.26).

B.1.2 Corrections to the Regime $\tau_C T \ll 1 - 2\lambda$

If, however, temperature is the smallest energy scale in the system, this means the two additional poles are sufficiently far from the $k = \pi$ point, we can use the method of intermediate asymptotics, where we do a perturbative expansion in small temperature around the pole at $z \approx 1$ in the first and fourth quadrants, up to the necessary accuracy. This gives the first term of $\mathcal{J}(\lambda)$ in Eq. (4.24b). For the other two poles, we can simply set $\omega \to 0$ in the denominator and pick up the other poles at

F. Stäbler, *Dissipation, Interactions, and Heat Transport in Quantum Hall Devices*, Springer Theses, https://doi.org/10.1007/978-3-031-90671-8

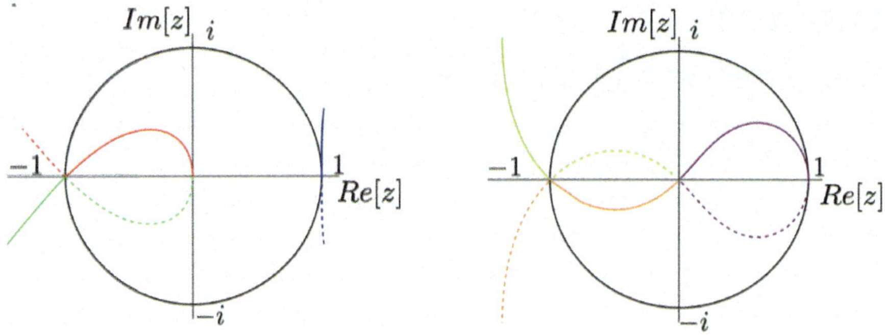

Fig. B.1 Poles of the heat integral in Eq. (4.23). The dashed lines indicate $\omega < 0$ and solid lines $\omega > 0$. The blue and purple poles are always present. In the non-interacting limit $\lambda \to 0$ we only pickup the purple pole. The other poles around $z = -1 \to k = \pi$ correspond to the second soft mode, the out-of-phase fluctuations of the collective mode. For finite λ these poles move towards the center of the contour, i.e. the momentum obtains a finite imaginary part. The excitation becomes gapped and we can rely on our low energy expansion, where the smallest energy scale is set by temperature, given by Eq. (4.24a)

$$z_{3/4} = -\frac{1}{2\lambda} + \sqrt{\frac{1}{4\lambda^2} - 1}. \tag{B.2}$$

Since the whole correlation function is already proportional to $\omega^2 \tau_C^2$ we immediately find the correction from this pole by setting $\omega = 0$ in the denominator. This gives the second term of $\mathcal{J}(\lambda)$ in Eq. 4.24b.

B.2 Edge States with Long Range Coulomb Cross Coupling

A loop of the length state of the edge L interacts with itself through a long-range Coulomb interaction with self-capacitance C. We introduce a non-local Coulombic cross coupling C_X between two of these mesoscopic capacitors which are separated by a distance W, see Fig. B.2. The Hamiltonian of the edge state inside the interaction region acquires the following additional terms.

$$\mathcal{H}_j = \frac{v_F}{4\pi} \int dx \left(\partial_x \phi_j(x,t)\right)^2 + \frac{Q_j^2(t)}{2C} + \frac{Q_j(t) Q_{j-1}(t)}{C_X}, \tag{B.3}$$

where $Q_j(t) = \frac{e}{2\pi} \int_0^L dx\, \partial_x \phi_j(x,t)$ is the integrated charge density of the loop and $j = 1, 2$ labels the left or right node.

The equation of motion of the bosonic field $\phi_j(x,t)$ is given by

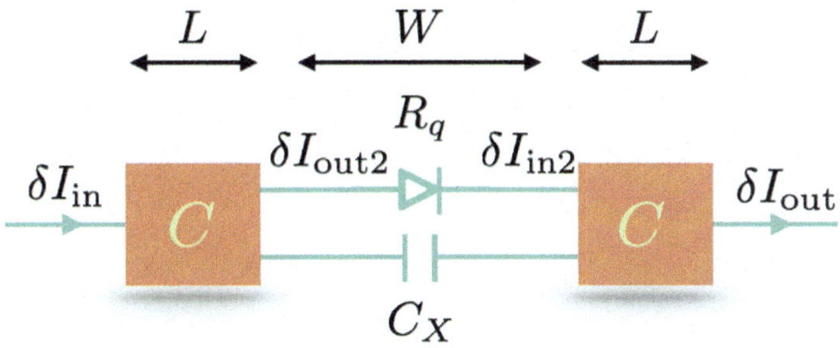

Fig. B.2 Cross capacitive coupling of mesoscopic capacitors

$$\partial_t\phi_j(x,t) + v_F\partial_x\phi_j(x,t) = -\frac{e}{C}\mathcal{Q}_j(t), \tag{B.4}$$

with $\mathcal{Q}_j(t) = Q_j(t) + \lambda\left(Q_{j-1}(t) + Q_{j+1}(t)\right)$, where we introduce the relative coupling strength $\lambda = \frac{C}{C_X} \in \{0, 1\}$ and $Q_0(t) = Q_3(t) = 0$. We assume the system is connected to a large reservoir without charge fluctuations and hence no capacitive cross couplings. The reservoir to the left has temperature T and emits the equilibrium current δI_{in}. We solve the equation of motion inside of the red interaction regions in terms of their boundary currents $\delta I_{\text{in}}, \delta I_{\text{in2}}, \delta I_{\text{out}}, \delta I_{\text{out2}}$, which allows us to express all currents as a function of only the equilibrium current δI_{in}. As expected we find that $\langle\delta I^2_{\text{out}}(t)\rangle = \langle\delta I^2_{\text{in}}(t)\rangle$. This is not necessarily the case for the intermediate current δI_{out2}. The Fourier transformation of the outgoing current $\mathbf{j}_{\text{out}} = (\delta I_{\text{out}}, \delta I_{\text{out2}})^T$ as a function of the incoming current $\mathbf{j}_{\text{in}} = (j_{\text{in}}, \delta I_{\text{in2}})^T$ is given by

$$\delta\mathbf{I}_{\text{out}} = \mathcal{S}\cdot\delta\mathbf{I}_{\text{in}}, \quad \mathcal{S} = \begin{pmatrix}\mathcal{A}(\omega) & \mathcal{B}(\omega)\\ \mathcal{B}(\omega) & \mathcal{A}(\omega)\end{pmatrix}, \tag{B.5a}$$

$$\mathcal{A} = \frac{-i\lambda\omega\tau_C x^2(\frac{\omega}{2})}{\left(1-\lambda^2-i\omega\tau_C\right)x^2(\frac{\omega}{2})+\omega\tau_C x(\omega)+\omega^2\tau_C^2 e^{-\frac{i\omega L}{v_F}}}, \tag{B.5b}$$

$$\mathcal{B} = \frac{\left(1-\lambda^2\right)x^2(\frac{\omega}{2})+\omega\tau_C x(\omega)+\omega^2\tau_C^2}{\left(1-\lambda^2-i\omega\tau_C\right)x^2(\frac{\omega}{2})+\omega\tau_C x(\omega)+\omega^2\tau_C^2 e^{-\frac{i\omega L}{v_F}}}, \tag{B.5c}$$

$$x(\omega) = 2\sin\left(\frac{\omega L}{v_F}\right). \tag{B.5d}$$

The scattering matrix $\mathcal{S}$ is unitary, but there is another channel where heat can propagate. With the additional constraint $\delta I_{\text{in2}} = \exp(i\omega W/v_F)\delta I_{\text{out2}}$, we can express the outgoing currents as a function of the boundary current δI_{in} only and compute their respective current- current correlation functions, i.e. the heat carried by the edge state. We analyze the non-trivial correlation function $\langle\delta I^2_{\text{out2}}(t)\rangle$ of the intermediate current at different limits of W and L.

B.2.1 Trivial Limits

We discover several trivial limits of the parameters L and W. In the limit $L \to 0$ there is no accumulation of charge in the capacitor and therefore no interaction between the reservoirs.

$$\lim_{L\to 0} \left\langle \delta I^2_{\text{out2}}(t) \right\rangle = J_q. \tag{B.6}$$

In the limit $W \to \infty$ the phase factors $\exp(\pm i\omega W/v_F)$ are fast oscillating, and we average them over a period. We do this by mapping the average integration to the unit circle contour C with the transformation $z \to \exp(i\omega W/v_F)$ and including only the residues inside the contour.

$$\left\langle \delta I^2_{\text{out2}}(\omega) \right\rangle = \oint_C \frac{idz}{2\pi} \frac{\left\langle \delta I^2_{\text{in}}(\omega) \right\rangle |\mathcal{B}(\omega)|^2}{(z - \mathcal{A}(-\omega))\,(z\mathcal{A}(\omega) - 1)}, \tag{B.7}$$

which evaluates to

$$\lim_{W\to\infty} \left\langle \delta I^2_{\text{out2}}(t) \right\rangle = J_q. \tag{B.8}$$

This means that the correlation between the outgoing and incoming intermediate currents is lost and the cross capacitive interaction does not influence the system and the current is equilibrium everywhere. For the rest of the paper we will thus consider the opposite limit of strong correlations between the intermediate currents and consider $W \to 0$.

B.2.2 Limit of $L \to \infty$

Similarly to the case of $W \to \infty$, we average the fast oscillations as a function of L by the following transformation $z \to \exp(i\omega L/v_F)$. This gives a slightly different integral compared to Eq. (B.7) of the form

$$\left\langle \delta I^2_{\text{out2}}(\omega) \right\rangle = \oint_C \frac{idz}{2\pi z} \frac{\left\langle \delta I^2_{\text{in}}(\omega) \right\rangle \mathcal{B}_z(\omega)\mathcal{B}_{\frac{1}{z}}(-\omega)}{\left(1 - \mathcal{A}_{\frac{1}{z}}(-\omega)\right)\left(\mathcal{A}_z(\omega) - 1\right)}, \tag{B.9}$$

where $\mathcal{A}_z(\omega)$ and $\mathcal{B}_z(\omega)$ is given by Eqs. (B.5b) and (B.5c) with the replacement $z \to \exp(i\omega L/v_F)$, note especially $x(\omega) = z + \frac{1}{z}$. $\mathcal{A}_{\frac{1}{z}}(-\omega)$ and $\mathcal{B}_{\frac{1}{z}}(-\omega)$ indicates the complex conjugate. The integral has three poles which are always inside the unit circle contour at

$$z_0 = 0, \tag{B.10a}$$

$$z_1 = \frac{\lambda^2 - 1 + i\lambda\omega\tau_C}{(\lambda-1)(1+\lambda+i\omega\tau_C)+i\omega\tau_C\sqrt{\lambda(\lambda+i\omega\tau_C)}}, \tag{B.10b}$$

$$z_2 = \frac{(\lambda-1)(1+\lambda+i\omega\tau_C)+i\omega\tau_C\sqrt{\lambda(\lambda+i\omega\tau_C)}}{\lambda^2 + i\lambda\omega\tau - (1+i\omega\tau)^2}. \tag{B.10c}$$

Picking up the residues of this pole gives the non-trivial result for the heat per frequency carried by the intermediate current.

$$\langle \delta I_{\text{out2}}^2(\omega)\rangle = f(\omega,\lambda)\langle \delta I_{\text{in}}^2(\omega)\rangle, \tag{B.11a}$$

$$f(\omega,\lambda) = \frac{\left(\lambda^2-1\right)^2 + \left(2+\lambda^2\right)\omega^2\tau_C^2 + \omega^4\tau_C^4}{\left(\lambda^2-1\right)^2 + (2+\lambda(3\lambda-4))\omega^2\tau_C^2 + \omega^4\tau_C^4}, \tag{B.11b}$$

where $f(\omega, \lambda)$ is the same correction as (4.11b), the one we found with reservoirs instead of mesoscopic capacitors.

B.2.2.1 Equivalence to the Reservoir Model

We note that the same equations Eqs. (B.11a) and (B.11b) can be found by replacing the large ($L \to \infty$) mesoscopic capacitors by ohmic reservoirs [1]. Note that the same corrections follow from solving the equations Eqs. (4.3a)–(4.3d) directly. The reason for this equivalence is the following. The current entering the mesoscopic capacitor propagating towards infinity loses the correlation with the current propagating from infinity towards the end of the capacitor. This means that an equilibrium source current, i.e. a Langevin source, satisfies this condition.

B.2.2.2 Finite *L* or *W*

In general finite L or W will introduce different kinds of oscillations of the heat carried by the intermediate edge state. This is natural and expected, since the negative drag effect arises from correlations between the charge in different nodes, which can be tuned by adjusting the retardation of the intermediate currents or the correlation of currents inside of the nodes themselves. This will lead to a modulation of the heat, but cannot change the sign of the correction to the heat flux.

Appendix C
Mesoscopic Multiplier and Fractionalizer

C.1 Dependence on External Resistance and Number of Loops

The heat flux transported inside of the loop depends on many parameters, namely, the external resistance, i.e. the number of channels attached to the Ohmic contact that do not form a loop, the length of the loop, and the number of loops. The Langevin equations for n external channels and m loops of length L are given by

$$-i\omega\Delta Q(\omega) = \sum_{k=1}^{n+m} \left(\delta I_{\mathrm{in},k}(\omega) - \delta I_{\mathrm{out},k}(\omega)\right), \tag{C.1}$$

$$\delta I_{\mathrm{out},k}(\omega) = \frac{1}{\tau_C}\Delta Q(\omega) + \delta I_{\mathrm{oc},k}(\omega), \tag{C.2}$$

$$\delta I_{\mathrm{in},k}(\omega) = e^{i\omega t_{\mathrm{fl}}}\delta I_{\mathrm{out},k}(\omega), \quad l > n \tag{C.3}$$

C.1.1 *Noise Power for Multiple Loops of Length L with Uniform Heating*

Let us assume equal temperatures for the noise powers of boundary currents and Langevin sources we find the noise power of one of the looped edge states to be $S_l(\omega) = f(\omega)S_{\mathrm{eq}}(\omega)$, with

$$f(\omega) = 1 - \frac{1}{m + \frac{2mn + n^2 + \omega^2\tau_C^2}{2(m - (m+n)\cos(\omega t_{\mathrm{fl}}) + \omega\tau_C\sin(\omega t_{\mathrm{fl}}))}}, \tag{C.4}$$

F. Stäbler, *Dissipation, Interactions, and Heat Transport in Quantum Hall Devices*, Springer Theses, https://doi.org/10.1007/978-3-031-90671-8

with $t_{\text{fl}} = L/v_F$. The heat flux carried by the looped edge state can be found by performing the integral in Eq. (2.26). If we take $L \to \infty$ a separate averaging of $f(\omega)$ over the oscillations shows that $f(\omega) \to 1$. From an experimental point of view the assumption of neglecting potential retardation effect due to a finite length of the loop is justified [2].

C.2 Scattering Matrix of the Capacitive Thermometer

The equation of motion for the capacitive thermometer is derived from a Hamiltonian of two counterpropagating edge states that interact capacitively within a region of size W with the coupling strength C_{th} (Fig. C.1).

$$\partial_t \phi_\sigma(x,t) + v_F \partial_x \phi_\sigma(x,t) = -\frac{e}{C_{\text{th}}} Q_{\text{th}} \theta(x)\theta(W-x), \tag{C.5}$$

where $Q_{\text{th}} = \sum_{\sigma'} \int_0^W dx \;\; \rho_{\sigma'}(x)$ is the total charge inside of the interaction region and $\sigma = \pm$ encodes the chirality of the edge states. The solutions of the equations of motion for $0 < x < W$ in terms of bosonic currents is given by

$$\delta I_\sigma(x,\omega) = \delta \tilde{I}_\sigma(\omega) e^{\frac{i\omega\sigma x}{v}} + \frac{1 - e^{\frac{i\omega\sigma x}{v}}}{i\omega\tau_{\text{th}}\sigma} \sum_{\sigma'} \int_0^W dx \partial_x \delta I_{\sigma'}(\omega, x), \tag{C.6}$$

to which we can assign the boundary currents

$$\delta I_{\text{in,th}}(\omega) \stackrel{!}{=} \delta I_+(0,\omega), \tag{C.7}$$

$$\delta I_{\text{out,th}}(\omega) \stackrel{!}{=} \delta I_+(W,\omega), \tag{C.8}$$

$$\delta I_{\text{in},n+1}(\omega) \stackrel{!}{=} -\delta I_-(0,\omega), \tag{C.9}$$

$$\delta I_{\text{out,th}}(\omega) \stackrel{!}{=} -\delta I_-(W,\omega), \tag{C.10}$$

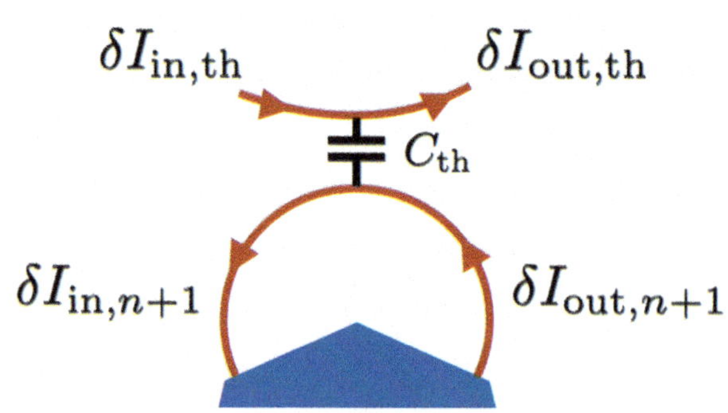

Fig. C.1 The labelling of currents of the thermometer

which allows us to solve for the S-matrix of the system which is given by

$$\begin{pmatrix} \delta I_{\text{out,th}} \\ \delta I_{\text{in},n+1} \end{pmatrix} = \begin{pmatrix} r_{\text{th}} & t_{\text{th}} \\ t_{\text{th}} & r_{\text{th}} \end{pmatrix} \begin{pmatrix} \delta I_{\text{in,th}} \\ \delta I_{\text{out},n+1} \end{pmatrix}, \tag{C.11}$$

where the transmission and reflection amplitudes of the thermometer are given by

$$t_{\text{th}} = \frac{1 - 2e^{i\omega t_W} + e^{2i\omega t_W}}{2 - 2e^{i\omega t_W} - i\omega\tau_{\text{th}}}, \tag{C.12}$$

$$r_{\text{th}} = \frac{1 - e^{i\omega t_W} + ie^{i\omega t_W}\omega\tau_{\text{th}}}{2 - 2e^{i\omega t_W} + i\omega\tau_{\text{th}}}, \tag{C.13}$$

where $t_W = W/v_F$ is the time of flight inside the interaction region and $\tau_{\text{th}} = R_q C_{\text{th}}$ is the RC-time of the capacitive interaction strength between the thermometer and the interacting mode.

C.3 Tunneling Hamiltonian Approach

The Hamiltonian of the system is given by

$$\mathcal{H} = \mathcal{H}_+ + \mathcal{H}_- + \mathcal{H}_T, \tag{C.14}$$

where

$$\mathcal{H}_\pm = \mp i v_F \int_{-\infty}^{\infty} dx\, \psi_\pm^\dagger(x,t)\partial_x\psi_\pm(x,t) \tag{C.15}$$

and

$$\mathcal{H}_T = \gamma\psi_+^\dagger(x_0,t)\psi_-(x_0,t) + \text{h.c.}, \tag{C.16}$$

where we use $\{\psi_+(x,t), \psi_+^\dagger(y,t)\} = \delta(x-y)$.

C.3.1 Average Heat Flux

The heat flux operator is given by

$$\begin{aligned} \hat{J}_{\text{tun}} &= -i\left[\mathcal{H}_+, \mathcal{H}_T\right] \\ &= -\gamma v_F \int_{-\infty}^{\infty} dx \left[\psi_+^\dagger(x,t)\partial_x\psi_+(x,t), \psi_+^\dagger(x_0,t)\psi_-(x_0,t) + \text{h.c.}\right], \end{aligned} \tag{C.17}$$

where we use the equation of motion and find

$$\left[\psi_+^\dagger(x,t)\partial_t\psi_+(x,t),\, \psi_+^\dagger(x_0,t)\psi_-(x_0,t)\right] = $$
$$= -\partial_t\psi_+^\dagger(x_0,t)\psi_-(x_0,t)\delta(x-x_0), \tag{C.18}$$

and

$$\left[\psi_+^\dagger(x,t)\partial_t\psi_+(x,t),\, \psi_-^\dagger(x_0,t)\psi_+(x_0,t)\right]$$
$$= -\psi_-^\dagger(x_0,t)\partial_t\psi_+(x_0,t)\delta(x-x_0). \tag{C.19}$$

Which yields

$$\hat{J}_{\text{tun}} = \gamma\left(\partial_t\psi_+^\dagger(x_0,t)\psi_-(x_0,t) + \text{h.c.}\right) \tag{C.20}$$

The average is given by

$$\begin{aligned}
J_{\text{tun}} = \left\langle \hat{J}\right\rangle &= -i\int_{-\infty}^{0} dt\,\langle[J(0),\mathcal{H}(t)]\rangle \\
&= -i\gamma^2\int_{-\infty}^{0} dt\left\langle\left[\partial_t\psi_+^\dagger(0)\psi_-(0) + \text{h.c.},\, \psi_+^\dagger(t)\psi_-(t) + \text{h.c.}\right]\right\rangle \\
&= -i\gamma^2\int_{-\infty}^{0} dt\left\langle\left[\partial_t\psi_+^\dagger(0)\psi_-(0),\, \psi_-^\dagger(t)\psi_+(t)\right]\right\rangle \\
&\quad + \left\langle\left[\psi_-^\dagger(0)\partial_t\psi_+(0),\, \psi_+^\dagger(t)\psi_-(t)\right]\right\rangle \\
&= -i\gamma^2\int_{-\infty}^{\infty} dt\left\langle\left[\partial_t\psi_+^\dagger(0)\psi_-(0),\, \psi_-^\dagger(t)\psi_+(t)\right]\right\rangle \\
&= -i\gamma^2\int_{-\infty}^{\infty} dt\left\langle\partial_t\psi_+^\dagger(0)\psi_+(t)\right\rangle\left\langle\psi_-(0)\psi_-^\dagger(t)\right\rangle \\
&\quad - \left\langle\psi_+(t)\partial_t\psi_+^\dagger(0)\right\rangle\left\langle\psi_-^\dagger(t)\psi_-(0)\right\rangle
\end{aligned} \tag{C.21}$$

C.3.2 Average Tunneling Current

$$\begin{aligned}
I_{\text{tun}} &= -i\int_{-\infty}^{\infty} dx\left[\psi_+^\dagger(x,t)\psi_+(x,t),\, \mathcal{H}_T\right] \\
&= -i\gamma\int_{-\infty}^{\infty} dx\left[\psi_+^\dagger(x,t)\psi_+(x,t),\, \psi_+^\dagger(x_0,t)\psi_-(x_0,t) + \text{h.c.}\right] \\
&= -i\gamma\int_{-\infty}^{\infty} dx\left[\psi_+^\dagger(x,t)\psi_+(x,t),\, \psi_+^\dagger(x_0,t)\psi_-(x_0,t)\right] \\
&\quad + \left[\psi_+^\dagger(x,t)\psi_+(x,t),\, \psi_-^\dagger(x_0,t)\psi_+(x_0,t)\right]
\end{aligned} \tag{C.22}$$

$$\left[\psi_+^\dagger(x,t)\psi_+(x,t),\, \psi_+^\dagger(x_0,t)\psi_-(x_0,t)\right] = \psi_+^\dagger(x,t)\psi_-(x_0,t)\delta(x-x_0) \quad (C.23)$$

$$\left[\psi_+^\dagger(x,t)\psi_+(x,t),\, \psi_-^\dagger(x_0,t)\psi_+(x_0,t)\right] = -\psi_+(x,t)\psi_-^\dagger(x_0,t)\delta(x-x_0) \quad (C.24)$$

$$I_{\text{tun}} = -i\gamma\left(\psi_+^\dagger(x_0,t)\psi_-(x_0,t) - \psi_-^\dagger(x_0,t)\psi_+(x_0,t)\right) \quad (C.25)$$

$$\begin{aligned} I = \langle I\rangle &= -i\int_{-\infty}^{0} dt\,\langle[I_T(0),\mathcal{H}_T]\rangle \\ &= -\gamma^2\int_{-\infty}^{0} dt\left\langle\left[\psi_+^\dagger(0)\psi_-(0),\,\psi_-^\dagger(t)\psi_+(t)\right]\right\rangle < \\ &\quad -\left\langle\left[\psi_-^\dagger(0)\psi_+(0),\,\psi_+^\dagger(t)\psi_-(t)\right]\right\rangle \\ &= -\gamma^2\int_{-\infty}^{\infty} dt\left\langle\psi_+^\dagger(0)\psi_+(t)\right\rangle\left\langle\psi_-(0)\psi_-^\dagger(t)\right\rangle < \\ &\quad -\left\langle\psi_+(t)\psi_+^\dagger(0)\right\rangle\left\langle\psi_-^\dagger(t)\psi_-(0)\right\rangle, \end{aligned} \quad (C.26)$$

C.3.3 Weak Backscattering Limit

In the weak back scattering limit the correlation functions of the "+" and "−" arm cannot be averaged separately. We thus start from the equation for current

$$I = \gamma^2\int_{-\infty}^{\infty} dt\left\langle\left[\psi_-^\dagger(t)\psi_+(t),\,\psi_+^\dagger(0)\psi_-(0)\right]\right\rangle \quad (C.27)$$

where we resum the exponentials due to the Gaussian nature of the theory. One of the terms in the tunneling current is given by

$$K(t) = \left\langle e^{-i\phi_-(t)}e^{i\phi_+(t)}e^{-i\phi_+(0)}e^{i\phi_-(0)}\right\rangle \quad (C.28)$$

$$-\ln K(t) = \int\frac{d\omega}{\omega}\left(1 - \frac{2(n+1)}{(n+1)^2+\omega^2\tau_C^2}\right)\frac{1-e^{-i\omega t}}{1-e^{-\omega/T}}, \quad (C.29)$$

which implies that the current, but also heat flux and Lorenz number can be obtained from the weak back scattering case by the mapping

$$n \to -(n+1), \tag{C.30}$$

which is a result of the duality of the weak back scattering and tunneling case.

C.4 Interacting Correlation Function

We would like to find the correlation function respecting the energy following separation of energy scales $T, t^{-1} \ll \tau_C \ll v_F a^{-1}$, where the time t is supposed to be far from the UV regime, i.e. we consider a long time limit. This gives

$$\begin{aligned}
-\ln C_-(t) &= \int \frac{d\omega}{\omega}\left(1+\frac{2n}{n^2+\omega^2\tau_C^2}\right)\frac{1-e^{-i\omega t}}{1-e^{-\omega/T}} \\
&= -\ln C_+(t) + \int_{\frac{1}{it}}^{\frac{v_F}{a}} \frac{d\omega}{\omega}\frac{2n}{n^2+\omega^2\tau_C^2} \\
&\quad + \frac{2}{n}\lim_{t\to\infty}\int_0^{\frac{v_F}{a}} \frac{d\omega}{\omega}\left(1-e^{-i\omega t}\right) - \log\left(i\frac{v_F}{a}t\right) \\
&\quad + \frac{2}{n}\sum_{n=1}^{\infty}\int_0^{\infty}\frac{d\omega}{\omega}(2-2\cos(\omega t))e^{-\omega n/T} \\
&= -\ln C_+(t) \\
&\quad + \frac{2}{n}\left(\log\left(int/\tau_C\right) + \gamma_{\mathrm{EM}} + \log\left(\frac{\sinh\left(\pi T t\right)}{\pi T t}\right)\right),
\end{aligned} \tag{C.31}$$

where $\gamma_{\mathrm{EM}} \approx 0.577\ldots$ is the Euler-Mascheroni constant. First note that the integral can be split into two parts: the free fermionic part and one part that contains the interaction. From now on we only consider the correction. We split the integral into two parts as before. For the temperature independent part, we cut the integral at $\omega \to \frac{1}{it}$, since the exponential function is fast oscillating for large t. Since the integral is in principle convergent for small frequencies, we have to take into account the possibility to find a constant term that does not vanish in the long time limit. To do this we also cut the integral at high frequencies only, subtract the logarithmic divergence, which is already accounted for in the long-time limit and take $t \to \infty$ in what remains. This gives an additional constant factor. The upper limit is given by the UV cutoff. For the temperature dependent part of the integral, we set $\omega_c \to \infty$, since the charging energy is much larger than the energy scale set by temperature and hence

$$\left\langle \psi_-^\dagger(t)\psi_-(t')\right\rangle = \underbrace{-\frac{iT}{2v_F}\frac{e^{i\mu_-(t-t')}}{\sinh\left(\pi T\left(t-t'-i\eta\right)\right)}}_{\text{free fermionic}}$$
$$\times\left(\frac{\pi\tau_C T}{ine^{\gamma_{\text{EM}}}}\frac{1}{\sinh\left(\pi T\left(t-t'-i\eta\right)\right)}\right)^{\frac{2}{n}}, \quad \text{(C.32)}$$

and for $\beta \to \infty$

$$\left\langle \psi_-^\dagger(t)\psi_-(t')\right\rangle = -\frac{i}{2\pi v_F}\frac{e^{i\mu_-(t-t')}}{t-t'-i\eta}\left(\frac{\tau_C}{ine^{\gamma_{\text{EM}}}}\frac{1}{t-t'-i\eta}\right)^{\frac{2}{n}}. \quad \text{(C.33)}$$

Note that the Euler constant appears only due to how the integral has been regularized. In this case we choose to cut-off the integral. Regularization with an exponential decay does not produce this constant as is shown for the free fermionic correlation function, but in any case it remains unphysical and should contribute to the normalization of the correlation function.

C.5 Average Current and Average Heat Flux

C.5.1 Average Current

The average current is given by Eq. (5.26), where the correlation functions are given by Eqs. (5.31) and (7.51), respectively. This yields the following integral

$$I_{\text{tun}} = \gamma^2\int_{-\infty}^{\infty} dt\,\Big(\left\langle \psi_+(t)\psi_+^\dagger(0)\right\rangle\left\langle \psi_-^\dagger(t)\psi_-(0)\right\rangle$$
$$-\left\langle \psi_+^\dagger(0)\psi_+(t)\right\rangle\left\langle \psi_-(0)\psi_-^\dagger(t)\right\rangle\Big). \quad \text{(C.34)}$$

At zero temperature, we note that the operator $\psi^\dagger$ applied to the ground state creates an electron-like excitation above the Fermi level (with the positive energy), while the operator ψ creates a hole-like one below the Fermi level (with the negative energy). One consequence of this is that all singularities in the first term are shifted to the upper half-plane of the complex variable t, whereas they are shifted to the lower half-plane in the second term. This means that only one term contributes depending of the sign of the bias, i.e. depending on if we close the contour in the upper and lower half plane. At finite temperatures, we also have occupied states above the Fermi level, because of thermal activation processes. We thus have to take into account both terms simultaneously.

$$I_{\text{tun}} = \left(\frac{\pi \tau_C T e^{-\gamma_{\text{EM}}}}{n}\right)^{\frac{2}{n}} \frac{\gamma^2 T}{4\pi v_F^2} \int_{-\infty}^{\infty} dt\, e^{i\frac{V}{\pi T}t} \sum_{\sigma=\pm 1} \frac{-\sigma}{(-i\sigma \sinh(t + i\sigma\pi T\eta))^{2+\frac{2}{n}}}, \tag{C.35}$$

where we shift the contour $t \to t + i\sigma\frac{\pi}{2}$ and use that $\sinh(t + i\sigma\frac{\pi}{2}) = i\sigma\cosh(t)$ and set $\eta \to 0$. This gives

$$I_{\text{tun}} = \left(\frac{\pi \tau_C T e^{-\gamma_{\text{EM}}}}{n}\right)^{\frac{2}{n}} \frac{\gamma^2 T}{4\pi v_F^2} \int_{-\infty}^{\infty} dt \frac{e^{i\frac{V}{\pi T}t}}{\cosh^{2+\frac{2}{n}}(t)} \sum_{\sigma=\pm 1} (-\sigma) e^{-\sigma\frac{V}{2T}}, \tag{C.36}$$

where we can perform the sum over σ, which gives

$$I_{\text{tun}} = \left(\frac{\pi \tau_C T e^{-\gamma_{\text{EM}}}}{n}\right)^{\frac{2}{n}} \frac{\gamma^2 T}{2\pi v_F^2} \sinh\left(\frac{eV}{2T}\right) \int_{-\infty}^{\infty} dt\, e^{i\frac{V}{\pi T}t} \frac{1}{\cosh^{2+\frac{2}{n}}(t)}, \tag{C.37}$$

where the last integral can be evaluate by taking $t \to \log(z)$ and express the result in terms of Gamma functions. We find

$$I_{\text{tun}} = \left(\frac{\pi \tau_C T e^{-\gamma_{\text{EM}}}}{n}\right)^{\frac{2}{n}} \frac{\gamma^2 T}{2\pi v_F^2} \sinh\left(\frac{eV}{2T}\right) \times \frac{2^{\frac{2+n}{n}} \Gamma\left(1 + \frac{1}{n} - \frac{ieV}{2\pi T}\right) \Gamma\left(1 + \frac{1}{n} + \frac{ieV}{2\pi T}\right)}{\Gamma\left(2 + \frac{2}{n}\right)}, \tag{C.38}$$

This gives the linear conductance.

$$G_n = \frac{\gamma^2}{2\pi v_F^2} \left(\frac{\pi \tau_C T e^{-\gamma_{\text{EM}}}}{n}\right)^{\frac{2}{n}} \frac{\sqrt{\pi}\Gamma\left(1 + \frac{1}{n}\right)}{2\Gamma\left(\frac{3}{2} + \frac{1}{n}\right)}, \tag{C.39}$$

which reproduces the free fermionic result

$$G_\infty = \frac{\gamma^2}{2\pi v_F^2}. \tag{C.40}$$

C.5.2 Average Heat Flux

The average heat flux is given by a similar expression.

$$J_{\text{tun}} = -i\gamma^2 \int_{-\infty}^{\infty} dt \left\langle \dot{\psi}_+^\dagger(0)\psi_+(t)\right\rangle \left\langle \psi_-(0)\psi_-^\dagger(t)\right\rangle - \left\langle \psi_+(t)\dot{\psi}_+^\dagger(0)\right\rangle \left\langle \psi_-^\dagger(t)\psi_-(0)\right\rangle, \quad \text{(C.41)}$$

in which the correlation functions depend on different temperatures and $\dot{\psi}(t) = \lim_{t'\to t} \partial_{t'}\psi(t')$. For equal temperatures the expression vanishes as expected from the second law of thermodynamics. We assume the temperature difference is small and expand in small δT, which gives the following heat flux.

$$J = -\frac{i\gamma^2}{16nv_F^2} T\delta T \left(\frac{\pi\tau_C T e^{-\gamma_{\text{EM}}}}{n}\right)^{\frac{2}{n}} \int_{-\infty}^{\infty} dt \sum_{\sigma=\pm1} (\sigma i)^{\frac{2}{n}} \times \frac{2t\left(1-n+\cosh\left(2t+2\pi Ti\sigma\eta\right)\right) - (2-n)\sinh\left(2t+2\pi Ti\sigma\eta\right)}{\sigma\sinh^4\left(t+\pi Ti\sigma\eta\right)\sinh^{\frac{2}{n}}\left(t+\pi Ti\sigma\eta\right)} \quad \text{(C.42)}$$

where we again shift the poles up and down similar to before and take $\eta \to 0$, which gives

$$\begin{aligned} J_{\text{tun}} &= \frac{\gamma^2}{8nv_F^2}\pi T\delta T \left(\frac{\pi\tau_C T e^{-\gamma_{\text{EM}}}}{n}\right)^{\frac{2}{n}} \int_{-\infty}^{\infty} dt\, \frac{1-n-\cosh(2t)}{\cosh^{4+\frac{2}{n}}(t)} \\ &= \frac{\gamma^2}{v_F^2}\pi T\delta T \left(\frac{\pi\tau_C T e^{-\gamma_{\text{EM}}}}{n}\right)^{\frac{2}{n}} \\ &\quad\times \frac{4^{\frac{1}{n}}\left((n-1)\Gamma\left(2+\frac{1}{n}\right)^2 - \Gamma\left(1+\frac{1}{n}\right)\Gamma\left(3+\frac{1}{n}\right)\right)}{n\Gamma\left(4+\frac{2}{n}\right)}, \end{aligned} \quad \text{(C.43)}$$

which can be evaluated as before by the transformation $t \to \log(z)$. Especially the non interacting case gives

$$J_\infty = \kappa_0 T\delta T \frac{\gamma^2}{v_F^2}. \quad \text{(C.44)}$$

The Lorenz number for a specific number of channels n is given by

$$\begin{aligned} \mathcal{L}_n &= \frac{J_n}{T\delta T G_n} \\ &= \frac{\left((n-1)\Gamma\left(2+\frac{1}{n}\right)^2 - \Gamma\left(1+\frac{1}{n}\right)\Gamma\left(3+\frac{1}{n}\right)\right)\Gamma\left(\frac{3}{2}+\frac{1}{n}\right)}{n\Gamma\left(4+\frac{2}{n}\right)\Gamma\left(1+\frac{1}{n}\right)} \frac{2^{2+\frac{2}{n}}\pi^{\frac{3}{2}}k_B^2}{e^2} \\ &= \frac{2+n}{2+3n}\frac{\pi^2 k_B^2}{e^2} = 3\frac{2+n}{2+3n}\mathcal{L}_\infty. \end{aligned} \quad \text{(C.45)}$$

Appendix D
Bosonization

D.1 Equality Between Fermionic and Bosonic Heat Currents

In this appendix, we show how the equilibrium heat current (2.25), given in terms of particle fluctuations described by a bosonic distribution function, can be derived from a description in terms of an electronic distribution function of chiral fermions. The latter quantity, given in Eq. (3.1), is repeated here for convenience as

$$J_{\text{ferm}} = \frac{1}{2\pi} \int d\epsilon (\epsilon - \mu) \left[f(\epsilon) - \theta(-\epsilon + \mu) \right] = \frac{\pi T^2}{12}. \tag{D.1}$$

The definition of the distribution function $f(\epsilon)$ is given

$$f(\epsilon) = \int dt e^{i\epsilon t} K(t), \tag{D.2a}$$

$$K(t) = \left\langle \psi^\dagger(t) \psi(0) \right\rangle, \tag{D.2b}$$

where $\psi^\dagger(t)$ is a fermionic operator creating a fermion at time t. In the time-representation, we express the fermionic heat flux (D.1), via Eqs. (D.2a) and (D.2b), as

$$J_{\text{ferm}} = -i\, \partial_t \left[K(t) - \frac{1}{2\pi i t} \right] \Bigg|_{t \to 0}, \tag{D.3}$$

where the second term in the brackets is the zero-temperature part. Next, we express the fermionic correlation function (D.2b) with the bosonization technique [3] as

$$K(t) = \frac{1}{2\pi a} \left\langle e^{-i\phi(t)} e^{i\phi(0)} \right\rangle = \frac{1}{2\pi a} e^{-\langle (\phi(t) - \phi(0)) \phi(0) \rangle}, \tag{D.4}$$

F. Stäbler, *Dissipation, Interactions, and Heat Transport in Quantum Hall Devices*, Springer Theses, https://doi.org/10.1007/978-3-031-90671-8

where $\phi(t)$ is a chiral bosonic field and a is a short-distance cutoff. The last equality in Eq. (D.4), that is, the formulation of the fermionic correlation function in terms of a bosonic correlation function, holds when the bosonic theory is Gaussian, which is true in equilibrium. Next, we use that the boson $\phi(t)$ is related to the current operator as

$$I(t) = -e\frac{\partial_t \phi(t)}{2\pi}, \tag{D.5}$$

and express the bosonic correlation function in term of the equilibrium charge noise power as

$$-\frac{e^2}{(2\pi)^2}\omega\omega'\langle\phi(\omega)\phi(\omega')\rangle = \langle\delta I(\omega)\delta I(\omega')\rangle \equiv 2\pi\delta(\omega+\omega')S(\omega), \tag{D.6}$$

where we used $\langle I(\omega)\rangle = \langle\phi(\omega)\rangle = 0$. Then, from Eq. (D.4), we have

$$\ln K(t) = -\ln(2\pi a) + \frac{2\pi}{e^2}\int\frac{d\omega}{\omega^2}S(\omega)(e^{-i\omega t} - 1). \tag{D.7}$$

Next, we identify $R_q = 2\pi/e^2$ and add and subtract a term $\ln(t)$, which allows us to write

$$\ln K(t) = -\ln(2\pi a) + \ln K_n(t) - \ln(t), \tag{D.8a}$$

$$\ln K_n(t) = R_q\int\frac{d\omega}{\omega^2}\left(S(\omega) - \frac{\omega\theta(\omega)}{R_q}\right)(e^{-i\omega t} - 1). \tag{D.8b}$$

By expanding $\ln K(t)$ to second order in time derivatives, dropping odd terms in the integrand, and inserting into Eq. (D.3), we finally arrive at

$$J_{\text{ferm}} = -\frac{1}{4\pi}\partial_t^2 \ln K_n(t)\bigg|_{t\to 0} = \frac{R_q}{4\pi}\int_{-\infty}^{\infty} d\omega\left[S(\omega) - \frac{\omega\theta(\omega)}{R_q}\right], \tag{D.9}$$

which is precisely Eq. (2.25). All the above steps can be repeated verbatim for the output heat current

$$J_{\text{out}} = \frac{1}{2\pi}\int d\epsilon(\epsilon - \mu)\left[f_{\text{out}}(\epsilon) - \theta(-\epsilon + \mu)\right], \tag{D.10}$$

which leads to Eq. (D.9), but with the replacement $S(\omega) \to S_{\text{out}}(\omega)$.

References

1. Slobodeniuk AO, Levkivskyi IP, Sukhorukov EV (2013) Equilibration of quantum Hall edge states by an Ohmic contact. Phys Rev B 88:165307. https://doi.org/10.1103/PhysRevB.88.165307
2. Sivre E, Anthore A, Parmentier FD, Cavanna A, Gennser U, Ouerghi A, Jin Y, Pierre F (2018) Heat Coulomb blockade of one ballistic channel. Nat Phys 14:145–148. https://doi.org/10.1038/nphys4280
3. Sukhorukov EV (2016) Scattering theory approach to Bosonization of non-equilibrium mesoscopic systems. Physica E 77:191–198. https://doi.org/10.1016/j.physe.2015.11.018

The manufacturer's authorised representative in the EU is Springer Nature Customer Service Centre GmbH, Europaplatz 3, 69115 Heidelberg, Germany. If you have any concerns regarding our products, please contact ProductSafety@springernature.com

Printed and bound by CPI Group (UK) Ltd, Croydon, CR0 4YY
30/09/2026
02225265-0010